SpringerBriefs in Pharmaceutical Science & Drug Development

For further volumes:
http://www.springer.com/series/10224

Aleš Prokop · Seth Michelson

Systems Biology in Biotech & Pharma

A Changing Paradigm

 Springer

Aleš Prokop
Chemical and Biomolecular Engineering
Vanderbilt University
VU Station B 351604
Nashville, TN 37235
USA
e-mail: ales.prokop@vanderbilt.edu

NanoDelivery International, s.r.o.
1327/113 Nádražní
69141 Břeclav-Poštorná
Czech Republic

Seth Michelson
Genomic Health Inc
101 Galveston Dr
Redwood, CA 94063
USA
e-mail: seth.michelson@comcast.net

ISSN 1864-8118
ISBN 978-94-007-2848-6
DOI 10.1007/978-94-007-2849-3
Springer Dordrecht Heidelberg London New York

e-ISSN 1864-8126
e-ISBN 978-94-007-2849-3

Library of Congress Control Number: 2011942347

Printed on acid-free paper

Springer is part of Springer Science+Business Media (www.springer.com)

Dedicated to my "teachers" and friends, the late Zdeněk Fencl (Prague), Arthur E. Humphrey, Elmer L. Gaden and Godfred E. Tong (USA) who influenced and shaped my professional career (AP)

Contents

Abbreviations

3D	Three-dimensional
3D/4D QSAR	Three dimensional/four dimensional QSAR
3D-QSAR	Three dimensional QSAR
ABC	Bayesian computation
ABM	Agent-based methods
ADMET	Absorption, distribution, metabolism, excretion, and toxicity
AMS	Accelerator mass spectrometry
ANN	Artificial Neural Networks
anti-CD40L	Antibody raised against CD40L region
ARACNE	Algorithm for the reconstruction of accurate cellular networks
BI	Bioinformatics
BLA	Biologic application
BOSS	Biological objective solution search
BSR	Biochemical system (network) reconstruction
CCA	Canonical correlation analysis
CD14-/-	Type of mice
CDD	Controlled drug delivery
CellML	Mark-up language
CG	Coarse-graining or CG computing
CNI	Correlation network inference
COAST	Complex automata for modeling and simulation of complex systems
COMBINE	Comparative binding energy
CSB	Computational systems biology
CSDD	Center for the study of drug development (Tufts)
DD	Drug discovery
DDD	Drug discovery and development

DDS	Drug delivery systems
DDv	Drug development
DOS	Diversity-oriented synthesis
dsRNA	Double strand RNA
EBI	European bioinformatics institute
ED	Enrichment designs
EGFR	Endothelial growth factor receptor
EPR	Passive uptake
ERK	Extracellular regulated kinase
FBA	Flux balance analysis
FBDD	Fragment based DD
FCR	Fluorochromatic reaction fluorescence
FDA	Food and drug association
FRET	Fluorescence resonance energy transfer
GNR	Gene regulatory network
GO	Gene ontology
hPXR	Humanized transgenic mice
HQSAR	Hologram quantitative structure-activity relationships
HT	High throughput
HTS	High throughput screening
iFBA	Integrated dynamic FBA
IFN-β	Interferon beta
IL-12	Interleukin 12
IL-15	Interleukin 15
JWS	Journal of web semantics
KEGG	Kyoto encyclopedia of genes and genomes
KNN	K-nearest neighbors
LDA	Linear discriminant analysis
MARS	Splines
MBDD	Model based drug design
MCA	Metabolic control analysis
MD	Molecular dynamics
MINDy	Modulator inference by network dynamics
miRNA	MicroRNA
MM	Molecular mechanics
MMR	DNA mismatch repair
MoA	Mode of action
MS	Multiple sclerosis
MVDA	Multivariate data analysis
NB	Naïve Bayes
NDA	New drug application
NME	New medical entity
NMR	Nuclear magnetic resonance

NOD	Nude mouse
OBRC	Online bioinformatics resources collection
ODE	Ordinary differential equation
OiCR	Ontario institute for cancer research
OMICS	Discipline of science and engineering for analyzing the interactions of biological information objects
PAT	Process analytical technology
PCA	Principal components analysis
PD	Parkinson disease
PD	Pharmacodynamics
PEGylation	Attachment of polyethylene glycol (PEG)
PEM	Protein epitope mimetic
PET	Positron emission tomography
PGN	Pharmacogenomics
PhRMA	Pharmaceutical research and manufacturers of America
PI3K	Phosphoinositide-3-kinase
PK	Pharmacokinetics
PKPD	Combined PK and PD
PLS	Partial least-squares
PM	Pharmacometrics
PromoLign	Simulation tool
PTEN	Phosphatase and tensin homolog
PupaSNP	Simulation tool
QbD	Quality by design
QSAR	Quantitative structure-activity relationship
R&D	Research and development
RO3	Rule of three
RA	Entelos rheumatoid arthritis
RAW	Mouse leukaemic monocyte macrophage cell line
RDD	Re-randomization design
ReguLign	Simulation tool
R-L	Receptor-ligand
RNAi	RNA interference
RNI	Reaction network inference
RNIDD	Reaction network inference for drug discovery
Ro5	Rule of five
ROT	Rule-of-thumb
RPART	Recursive partitioning and regression trees
SAR	QSAR
SB	Systems biology
SBML	Systems biology markup language
siRNA	Small nterfering RNA
SG	Systems genetics

SNP	Single nucleotide polymorphism
SSM	Scale separation map
SVM	Support vector machines
TGF-β	Transformation growth factor beta
TNF-α	Tumor necrosis factor alfa
uHTS	Ultra high throughput screening
WW2	Second world war

Acknowledgments

The authors appreciate a critique by Béla Csukás (Kaposvar, Hungary).

Abstract

Systems Biology (SB) is suite of technologies and methodologies that resulted, conceptually, from the merging of two basic paradigms, reductionism and holism. It represents a combination of reductionist and holistic approaches to the relationships among the elements of a system, with the goal of identifying its emergent properties and defining, quantitatively, molecular, cellular, tissue, organ and whole body processes. One manifestation of SB is as a tool for hypothesis generation about a system that is typically too large and complex to understand by simple reasoning.

The US is currently well ahead of the rest of the world in the development and application of SB and its principles especially as they pertain to basic medical research and development. This lead is largely due to an earlier start in the academic arena (7–9 years ago in US vs. 4–5 years elsewhere; Rubenstein 2008). However, there is evidence of rapid development in both the UK/EU and Japan and the gap is narrowing, particularly in UK. From an industrial point of view, the Pharmaceutical Industry based in the US and UK can capitalize on these opportunities and gain the benefits of this technology. Early and sustained investments in SB and its enabling technologies will likely produce financial rewards for any pharma company so inclined. Many big Pharma companies have already invested in SB and Bioinformatics (BI) (for definitions of SB and BI see Chap. 1) and are set to maintain their lead. The industrial significance of SB is thus clear.

This review intends to educate a large population of cell and molecular biologists in the use of the quantitative tools that are available to them to solve the critical problems they face. Many educational institutions (and particularly their medical divisions) at present are heavily business-oriented realize that in this particular industrial environment that dollar counts. It is thus important that biologists recognize early in their research the utility of SB and how this approach can help to generate new therapeutic leads and substances useful for human health. Educational curricula in the life sciences have typically been based on the atomistic belief that one can decompose complex systems into their components and that a detailed investigation of each these components individually will in itself lead to novel biological insights. Indeed this is true in numerous instances. However, increasing acknowledgment of the importance of studying whole systems, as well as their components, has led to an emphasis on teaching not just a reductionist view of biology, but also a complementary constructionist view. {Note: Bioprocess engineering issues, as related to a systems approach at manufacturing, are not included in this review. However, included in our broader definition, we do incorporate some of the issues surrounding BI in this review}. Overall, we attempt to answer a question: Can Systems Biology deliver on academic funding and business profits?

Chapter 1
Introduction: Discovery and Development—New Facet of Industry, New Tools and Lead Optimization

In the face of the challenges associated with expiring drug patents, the rising cost of R&D, and payer pressure on pricing, most major pharmaceutical companies are seeking ways to enhance productivity, reduce costs and augment their late-stage pipelines. Recent technological applications have witnessed the development of data-rich, genome-scale functional screens, large collections of reagents and protein microarrays, and the addition of databases and algorithms for data mining. Systems biology takes advantage of these technological breakthroughs to represent a combination of reductionist and holistic approaches to the relationships among the elements of a system. The obvious goal of this effort is to identify the emergent properties of a system, and define, quantitatively, molecular, cellular, tissue, organ and whole body processes that best lend themselves to external manipulation. The emergent properties of complex systems are recognized in terms of species and topologies, teaching us that one should target network states (resulting from dynamic molecular and physiological networks and their emergent behavior) and not just individual genes or proteins. While emergent properties are quite common in science and engineering, they are rarely exploited in biology and medical R&D.

Definitions:

- *In silico* **biology** (also systems biology) is an expression used to mean "performed on a computer or via computer simulation. The phrase refers to 'experiments' done outside living organisms".
- **Multiscale modeling** studies properties of a system over a wide range of length and time-scales and hierarchies. Multiscale models include components from two or more levels of organization (multiple length scales) or multiple time scales.

A. Prokop and S. Michelson, *Systems Biology in Biotech & Pharma*,
SpringerBriefs in Pharmaceutical Science & Drug Development,
DOI: 10.1007/978-94-007-2849-3_1, © The Author(s) 2012

- **Systems biology** (SB) employs a rational approach, a mix of analytical and synthetic routes to delineate emergent properties relevant to higher hierarchy levels, aiming to explain and predict, quantitatively, molecular, cellular, tissue, organ and whole-body processes.
- **Empirical approach**: the prevalent trial and error approach, a highly inefficient approach
- **Bioinformatics** (BI) the discipline that deals with the computational needs of biology to decipher the genetic blue-print and to infer the structure and function of gene-encoded proteins and RNAs, including data acquisition, storage, organization, archival, analyses and/or visualizations. It is a very computationally-intensive field. Practical applications include design of biomimetic proteins, quantitative diagnosis and the prognosis of diseases.
- **Drug Discovery and Development** (DDD), a collective term for *drug discovery (DD)* and *drug development (DDv)*.

1.1 Scope and Content of this Review

One of the most critical share-price drivers for drug companies is their ability to innovate and create new patent-protected products. Unfortunately, the industry is facing a serious R&D productivity problem. Most big pharma companies spend an average of over 15% of sales on R&D, but recent years have seen diminishing returns on this investment. The R&D productivity problem is exacerbated by the rapidly increasing cost of R&D—in 2001 the Tufts Center for the Study of Drug Development (Tufts CSDD) put the cost of developing a new drug and bringing it to market at US $802 million. In 2006, Tufts CSDD estimated the average cost of developing a new biotechnology product to be US $1.2 billion. The reasons for, and extent of, the R&D productivity crisis, is controversial because of the different metrics used and the problems associated with these as an accurate measure of performance, given the long timeframe associated with developing and bringing a product to market. Part of the problem is the need for larger and more complex clinical trials, as well as expensive new enabling technologies. The mega-mergers of the late 1990s promised that greater scale in R&D would increase the probability of successful product development with 'more shots on goal', but in reality this has not translated into improved pipelines. As a consequence there is a lack of genuinely innovative drugs to replace revenues lost through patent expiry; many new drug launches are 'me-toos'. Few in the industry would deny that it is facing a serious problem in generating new products to replace those lost to generic competition or product withdrawals.

One commonly-quoted piece of evidence for the R&D productivity crisis is illustrated in Fig. 1.1 (¹PhRMA Pharmaceutical Industry Profile 2007, FDA) which compares R&D expenditure by the major players versus the number of new

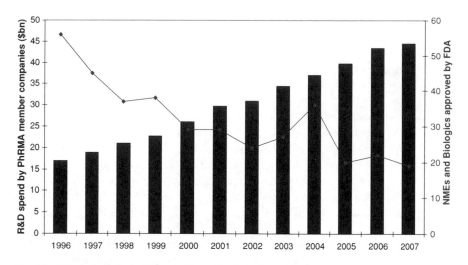

Fig. 1.1   The industry's declining R&D productivity is demonstrated by the declining number of approved products, despite a huge increase in R&D investment

products approved annually by the FDA. These have been moving in opposite directions since the mid-1990s, with fewer approved products every year despite large increases in R&D spend. The latest figures, released by the FDA and the US industry trade body PhRMA, show no change in this trend, with PhRMA members' R&D expenditure reaching an all-time high of US $44.5 billion in 2007. Additionally, the global industry spent an estimated US $58.8 billion *en toto*, even as the FDA approved just 19 NMEs and BLAs—the lowest number in a generation.

Even the number of new biologics reaching the regulatory submission stage in the US (as measured by the number of BLAs submitted to the FDA) has been slowing down over the past 15 years, indicating that R&D productivity issues affect both chemical-based and biological-based therapies.

On the other hand, recent technological advances have witnessed the development of genome-scale functional screens, accumulation and validation of large collections of reagents, and development of protein microarrays, databases, and algorithms for data and text mining. Taken together, these newer technologies enable unprecedented descriptions of complex biological systems, which are testable by mathematical modeling and simulation. While the methods and tools are advancing, it is their iterative and combinatorial application that defines the SB approach.

SB represents a combination of analytical and synthetic approaches to uncovering and quantifying the relationships between the elements of a biological system, with the ultimate goal being a clear understanding of that system's emergent properties. SB seeks to describe all of the elements of the system, define the biological networks that inter-relate the elements of the system to each other, and finally to characterize the flow of information that links these elements and their networks. The resultant structure will describe and characterize the emergent biological properties and processes of that system, which include molecular, cellular, tissue, organ and whole body interactions.

SB aims to understand the operation of complex biological systems at a fundamental level and then, based on these understandings ultimately develop predictive models of human disease. A specialized branch of SB, dedicated to the quantitative aspects of these efforts, is called *computational systems biology* (CSB). Although meaningful molecular level models of human cell and tissue function are a distant goal, SB efforts are already influencing drug discovery (DD). Large-scale gene, protein and metabolite measurements (the *OMICs*) dramatically accelerate hypothesis generation and testing in disease models. The latter integrate knowledge of organ and system-level responses into a single unified context and help to prioritize targets and to design clinical trials.

In addition, recent progress in CSB has had an impact on DD applications. The focus is on novel *in silico* methods to reconstruct regulatory networks, signaling cascades, and metabolic pathways, with an emphasis on comparative genomics and micro-array-based approaches. Promising methods, such as the mathematical simulation of pathway dynamics has proven useful during Drug Discovery, especially with respect to target identification, target validation, and optimal experiment design.

Coupling mathematical models across large ranges of length and timescales is central to describing complex behaviors in multifaceted systems and is, therefore, fundamental to making systems biology work. Such coupling may be performed by means of hierarchical and hybrid multiscale modeling. "Hybrid" is meant to represent both the merging continuous and discrete models, as well as quantitative and qualitative results. These modeling methods are expected to bring numerous benefits to biological and medical research, as the properties of a system may be studied over a wider range of length and timescales, a key aim of SB. Multiscale models couple behavior at the molecular biological level to that at the cellular level, thereby providing a route for calculating many unknown parameters. It is equally true that this multiscale environment enables investigation of effects that might be expressed at the cellular level as they may be induced by small changes at the biomolecular level. Such small changes might include, but are certainly not limited to induction of a genetic mutation or the presence of a drug. The modeling and simulation of biomolecular systems is itself very computationally intensive. Hybrid models with their associated algorithms can point the way towards the integration of molecular and more granular representations of matter.

SB is already being practiced in Pharma, and has come to represent an encompassing tool for coupling the reductionist and holistic approaches to DDD into a single unified conceptual structure. This report will discuss recent developments in both approaches and how they can be integrated into a single means for improving the performance of the industry. It will largely focus on the quantitative aspects and tools available to the biomedical researcher, and may include some qualitative approaches that may be equally useful in DDD.

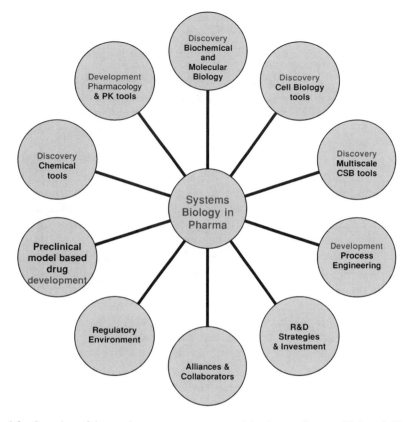

Fig. 1.2 Overview of the employment systems approach in pharma. Systems Biology in Pharma is attempting to rationalize the DDD pathway by developing an integrated suite of qualitative and quantitative rules. The goal is to change the current R&D paradigm and speed up the entire process of bringing a medicine to market. Covered subjects in R&D (process development, regulatory, R&D strategies, ethical issues and public–private partnership are not covered in this review)

Because of the huge literature of scientific papers on SB, this report will limit the scope of our examples to those that have been approved for clinical use, or are currently in clinical trials. We will also cite new, basic approaches which may change this field in a near future. Even so, it is impossible to know of, and include, all such examples and to properly credit all the key people who helped to bring the various technologies and ideas to the clinic. The authors apologize in advance for all omissions.

The systems approach, the umbrella methodology, emphasizes two sets of tools: qualitative and quantitative, both equally important. Some qualitative tools (heuristics) often substitute for the lack of quantitative tools and provide an order of input for significant design criteria for decision makers involved in the R&D project portfolio. The graphical overview of this structure is shown in Fig. 1.2.

Soft issues in R&D (those outside the purely scientific realm) are complementary to the hard/scientific skill set required for systematic investigation. However, these issues may be decisive in the overall process of putting a drug on the market, potentially overriding any number technical issues in their impact. These "soft" aspects of the business could be included in a qualitative reasoning simulation model, augmenting the hard science in a decision analytic context.

This report examines the status of SB within big pharma and the biotech industry and looks at the different system tools available at each level of hierarchy, while noting that the virtual chemical space does not belong to any kind of structural biological hierarchy and the biochemical and molecular space is, of course, part of the sub-cellular level. The report updates the drug discovery and drug development (DDD) tools listed by the author in his previous review of the industry [1, 2]. It identifies new areas of DD and appropriate associated tools, paying special attention to quantitative approaches (*in silico*), which are now part of this more global systems view.

The report does not provide a comprehensive list of approaches/tools as only those considered relevant and with development potential are reviewed. The same applies for the most recent iterations of each tool; they are described because they play a significant role in helping to shape the systems view.

This report brings, for the first time, attention to the importance of the systemic/holistic view to emphasize the two-directional pathway of the biological systems approach in the pharmaceutical industry and its potential. It *emphasizes two sets of tools: qualitative and quantitative*, both equally important. Some qualitative tools (heuristics) often substitute for the lack of quantitative tools and provide an order of input for significant design criteria for decision makers involved in the R&D project portfolio.

1.2 Drug Discovery and Development

DD usually includes: conceptualization, determination of (hypothetical) mechanism of effect, verification through assay (including assay development and design), screening of compounds against the purported target, hit identification of potential compounds, target validation, lead demonstration, and lead optimization. It usually also involves a limited in vivo proof of concept in animals and concomitant demonstration of a possible therapeutic index (preclinical pharmacology and safety/toxicology). Drug Development (DDv) begins when a decision is made to put a molecule into phase I clinical trials. This report uses the collective term Drug Discovery and Development (DDD) to encapsulate the entire evolution of a medicine. DDD often features the following:

- The time from concept and patent to approval of a new drug entity is commonly listed as 8–10 years but may typically be 15–17 years.
- The vast majority of molecules fail during discovery and development.

The estimated cost of bringing a successful product to market is very high. The amortized cost of approximately US $1 billion is commonly used, but the figure for a major drug may now be even higher.

Development involves various phases that can be grouped in three stages:

- Pre-clinical investigation and characterization
- Clinical trials
- Marketing (or post-approval)

Most large Pharma companies usually have a number of compounds in their pipelines. The drug pipeline is an important evaluator of future prospects of a company. Usually the more compounds in the pipeline the better company is. Assessing risk and eliminating compounds that may not eventually get approved is essential.

The drugs themselves have a life cycle, beginning at the earliest stage of discovery and leading through the development stages, regulatory review, market authorization, and post-market activities. The market life cycle can be grossly categorized into four stages:

- Introduction
- Growth
- Maturity and
- Decline

while brand life span can be divided into three stages:

- Launch
- Maintenance
- Retirement

Patent expiration is another impediment to successful brand sales. From 19% in 1984, generics' prescription share of the market rose to more than 50% in 2004.

1.3 Defining SB: Key Components

We can define Systems Biology as a bi-directional process, linking integrative, top-down, mechanism-based or deductive modeling with bottom-up hypothesis-driven or inductive modeling. Only when inductive logic (bottoms-up model) is iteratively linked to deductive reasoning (Mechanism-based) does true feedback and learning accrue. Therefore, it is often necessary to find a middle ground methodology to systematically study complex biological and biomedical processes in the context of their experimental data. One way to do that is by developing a set of tools (both qualitative and quantitative) to evaluate relationships and interactions among various system components under multiscale and dynamic conditions. The work product of this effort is typically an integrated phenotype at the whole body level, which will, by definition, include lower level hierarchies, at molecular, cellular, tissue and organ levels. In a clear departure from the mechanism-based reductionist approach, SB embraces both arms of scientific dogma, reductionist and holistic, by employing an integrated, closed-loop learning middle ground. As more SB global tools are developed and better extraction of emergent properties is instituted, Biomedical R&D will experience a paradigm shift to this more systemic approach. By closing the loop between induction and deduction in a systematic, modeling context, the field will gain a deeper understanding of disease mechanisms (including the upper level interactions), and derive benefits for expedited DD and improved drug safety and efficacy profiles.

The *in silico* method (which SB mostly employs), refers to the fact that the 'experiment' is conducted within simulation software, rather than in vitro (i.e., in a test tube or Petri dish), or in vivo (an animal model of human disease or more spectacularly in the human subject him/herself). When combined with traditional methods, this approach may allow molecules to fail early rather than late in the DDv path, where the real loss is typically accrued. By employing the principles of closed loop learning encompassed by SB, models of cell systems are produced that can be used to analyze pathways in disease as opposed to normal conditions, enabling an understanding of the effects of potential drugs on pathways. Virtual experiments can then be run on systems that are both dynamic and can be customized for different cell types. These system come pre-packaged with research-based protocols that can be tweaked at various levels for in depth analysis and research.

This approach enables researchers to qualify the biological target and potentially any associated toxicity at the nascent stage of discovery itself, producing tremendous savings in time. In the highly competitive pharma industry, every year saved in drug research could translate into many hundreds of millions of dollars in realized revenues, besides giving a first-mover advantage in the market. To create this type of functionality, one must compile a competent interdisciplinary team of experts in biology, mathematics, chemistry and computer science. But if it were really that easy, anyone would do it. It is a lot more nuanced and difficult than that. However, given that the prevalent trial and error approach has a demonstrated track record of being time consuming and inefficient, adopting technologies such as these may be the best way forward. But it should also be noted that while SB may improve the researcher's understanding of human physiology and disease, drug discovery remains a very slippery path.

1.4 A Brief History of Systems Biology (SB): In Terms of Key Advances

- Weiss in 1924 introduced the term 'systems biology'. Wiener in 1948 established cybernetics and reinforced the feedback concept; earlier, Bernard (in the 19th century) and then Cannon in 1932 determined the importance of negative feedback for maintaining homeostasis; molecular feedback in bacteria was discovered by Yates and Pardee in 1956.

- First numerical simulation in biology (neural cell action potential) was published in 1952 by Hodgkins and Huxley.
- Quantitative enzyme kinetics flourished between 1900 and 1970.
- Noble in 1960 developed a computer model of a heart pacemaker.
- Biochemists (late 1960s) developed biochemical systems theory.
- Mesarovic in 1966 established the formal discipline of systems biology.
- Polyani in 1966 recognized that upper level behavior requires the lower level behaviors.
- Kaczer and Burns established Metabolic Control Theory, 1973.
- Von Bertalanffy proposed General Systems Theory, 1976.
- The birth of functional genomics is marked for the 1990s, followed by bioinformatics development.
- Several microorganisms, mostly pathogens, were sequenced (1994–1998).
- Draft of the human genome sequence produced, 2000.
- Institute of Systems Biology was established in Seattle and Tokyo (2001).
- The first version of the Systems Biology Markup Language (SBML), used to encode biological reaction sets, was released in 2001.
- Kitano in 2002 defined Computational Systems Biology.
- Steuer (2003) and Morel (2004) defined metabolic correlation network^{TM}.
- Barabasi defined network biology, while defining formal properties of networks, 2004.
- Westerhoff and Palsson in 2004 defined systems biology as a mixed reductionist and holistic approach.
- Barabasi in 2007 defines disease as a network disorder.

1.5 Perspective and Potential Impact of Systems Biology on Academic Funding and Pharma R&D and Cost Savings

The following scheme of the SB paradigm (Fig. 1.3) is proposed as a modification of the already existing standard R&D flow chart (not shown), with a detailed discussion of the particular tools discussed later in the appropriate sections of this Report.

The **Systems Biology cycle** requires a careful set-up of experimental cycles enriched with quantitative tools. The inductive-deductive feedback loop consists of perturbation (experimental design and assay development), measurement, modeling, hypothesis generation, and feedback to perturbation, etc. This is a standard science-learning cycle process. The precise order of the steps of the cycle may vary depending on which facts are known, which observations can be made, the strength of the evidence that backs up inferences that can lead to a robust hypotheses (or educated guesses), and which components of the hypothesis can be tested (i.e., which technology may or may not work in a particular assay design).

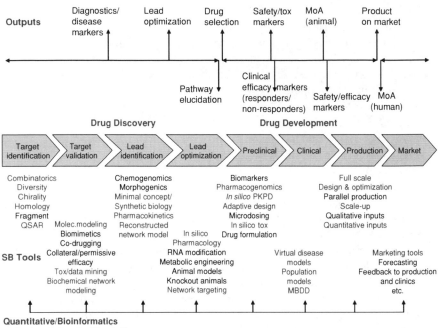

Fig. 1.3 Systems Biology Paradigm No.1. Although the new pharma paradigm, enriched with systems tools (enhanced DDD, EDDD), is depicted above as a linear process, overlaps, feedbacks (*recycles*) and detours are often encountered. Along the path, the amount of data, complexity and cost increases. Note that the SB tools listed for each DDD stage are not necessarily in the order of sequence of the sections of this report. Bioinformatics inputs are listed in *pink*

Key design features of this Review: We will list all salient features of each hierarchical level, starting with discovery, and discuss both qualitative and quantitative phenomena observed therein; for those where quantitative tools are available, they are underlined, their availability listed, and the state of the art indicated. Both qualitative and quantitative modelling are discussed as part of systems view, as the systems approach calls for coverage of all systemic phenomena, regardless of their computational status.

References

1. Prokop A (1995a) Challenges in commercial biotechnology. In: Laskin A (ed) Part I: product, process and market discovery. Adv Appl Microbiol 40:95–154
2. Prokop A (1995b) Challenges in commercial biotechnology. In: Laskin A (ed) Part II: product, process and market discovery. Adv Appl Microbiol 40:155–236

Chapter 2
Discovery: Use of Systems Biology for Identifying Targets

In our introduction, we emphasized that a combination of reductionist (mechanism-based) and holistic (hypothesis-based) tools in the drug screening process may increase the efficiency of overall Drug Discovery. Among notable holistic tools are screens that target discovery and characterization of molecular probes (compounds) that will enable the investigation of fundamental biological function at molecular, cellular and whole organism levels. Such screening usually occurs at the earlier stages of drug discovery. However, sometimes the required biological insights and understanding of both the normal physiology and the aetiology of the disease process are too sparse to allow for an appropriate screening effort. In those instances, exploratory experiments and ad hoc assay designs are required to test and refine particular hypotheses regarding potential target viability. To that end we discuss other holistic tools aimed at characterizing biological dynamics at the biochemical and molecular biology levels in Chaps. 3 and 4.

An analysis of the druggable space suggests that pharma is reaching a saturation point in terms of available targets, and mechanism-based screens may not represent an optimal strategy for targeting this very narrow point within the druggable space. One hole in this technological strategy is that it avoids cross fertilization between disease-modifying genes and the druggable genome [1]. Additionally, drug pleiotropy (Chap. 4) may cause undesirable off-target effects. A rationale for selecting drug combinations by systems tools is proposed in Chap. 3.

Combinatorial chemistry tools help to rationalize the DD process by exploiting several chemical and informational tools (none of them being relevant to a systems approach described in this report, however). The Achilles heel of the present approach is a strong emphasis on high-affinity ligands (low affinity ligands are important for cell–cell interactions and the immune system) and the failure to detect molecules that bind covalently to the receptor.

Qualitative and quantitative screens and other filters only facilitate the identification of lead substances (i.e., those compounds that will lend themselves to optimization by classical medicinal chemistry efforts). Since every hit will not

A. Prokop and S. Michelson, *Systems Biology in Biotech & Pharma*,
SpringerBriefs in Pharmaceutical Science & Drug Development,
DOI: 10.1007/978-94-007-2849-3_2, © The Author(s) 2012

necessarily translate into a marketable product, the aim of this stage of the search is to seek a facility that will eventually enable virtual chemistry screening.

Definitions:

- **Top-down** is a reductionist approach analyzing a system down to the mechanistic level.
- **Bottom-up** is a holistic approach, seeking to cross boundaries and hierarchic organization to synthesize the knowledge to the cell, organ and organismic levels. This is a novel approach, building up from mechanistic knowledge.
- **Druggable** genome represents a part of the genome which could potentially be accessed by an effective drug candidate. It represents amenability to modulation of a target by drugs.
- **MoA** (Mechanism of Action) refers to the specific biochemical interaction through which a drug substance produces its pharmacological effect.
- **SAR (QSAR)** is the process by which chemical structure is (quantitatively) correlated with a well defined process, such as biological activity or chemical reactivity.

This part encompasses a short introduction to Chaps. 2 and 3, and covers target-based and physiology-based approaches as well as combined systems approaches. In a target or mechanism-based (top-down, or so-called reductionist) approach, scientists first identify a protein or molecular mechanism relevant to a disease process and then use screening to find compounds that interact with or modulate it. The present effort may have swung too far in the direction of the reductionism implicit in target and mechanism-based drug discovery.

Bottom-up (holistic) methods of identifying interesting targets and the compounds that interact with them, often result in compounds that, in animal systems or in humans, show that the underlying mechanism is more complicated than was previously thought. This is an argument for less reductionism. On the other hand, earlier, a lot of DD was driven by phenotypic screening, in which compounds were tested for a desired readout in cell or animal systems and their target and mechanism were typically unknown at first. This is called a phenotypic cell and organism-based (holistic) approach.

The mechanism-based approach is unlikely to be abandoned anytime soon, but some people believe that the traditional cell- and organism-based strategy (also called the physiological strategy) should not have been discarded completely and should now be given renewed emphasis. *The combined reductionism and holistic approach is most germane to the systems approach*, utilizing both directions in the feedback loops [2].

2.1 Identifying Targets and Druggability Space

Since the early 1990s, the target-based drug discovery paradigm has been the dominant approach in the pharma industry. It takes a rational approach by defining the specific molecular mechanism or Mechanism of action (MoA) to be targeted by the treatment from biological and clinical findings. However, it does not translate

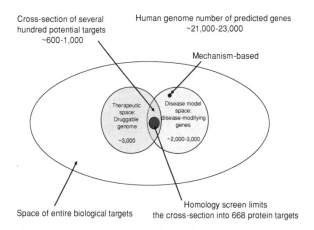

Cross-section of several
hundred potential targets
~600-1,000

Human genome number of predicted genes
~21,000-23,000

Mechanism-based

Therapeutic
space:
Druggable
genome

~3,000

Disease model
space:
disease-modifying
genes

~2,000-3,000

Space of entire biological targets

Homology screen limits
the cross-section into 668 protein targets

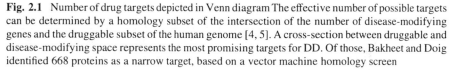

Fig. 2.1 Number of drug targets depicted in Venn diagram The effective number of possible targets can be determined by a homology subset of the intersection of the number of disease-modifying genes and the druggable subset of the human genome [4, 5]. A cross-section between druggable and disease-modifying space represents the most promising targets for DD. Of those, Bakheet and Doig identified 668 proteins as a narrow target, based on a vector machine homology screen

into a high success rate for novel targets, presumably because our level of insight into disease dynamics and their associated biological processes is not sufficient to predict the therapeutic value and druggability of a novel target. Figure 2.1 shows a logical Venn diagram depicting how targets can be classified and/or eliminated from a potential search. Note, the Bakheet and Doig estimate may look very pessimistic. However, targeting proteins is likely a better approach since nearly all drugs act at the protein level (except transcription process inhibitors). Fortunately, the proteome is a much better target than the genome, owing to alternative splicing and post-transla-tional modifications. Mechanism-based approaches select a single known mecha-nism and are represented by a single Point; it limits the effort to an unreasonably small sampling of target space. Physiology-based and function-based approaches do not employ any assumption regarding the MoA and operate within their respective disease-modifying spaces. Disease-modifying genes are those involved in disease onset or progression. More on global mapping of pharma space is in Paolini et al. [3].

   The characteristics of druggable compounds should be such that they can be delivered effectively to the target tissues and cells [6]. These authors estimated that the upper limit of molecular targets in the human genome that represent an opportunity for further therapeutic treatment is approximately 6,300 human pro-teins that are similar to sequences of known protein targets collected from the DrugBank database (about 20% of human proteome). On the other hand, Drews [7, 8] identified about 5,000–10,000 potential targets on the basis of number of disease-related genes. Theoretical prediction of protein function from the gene sequence computationally remains an excruciatingly imprecise science. Despite these hurdles, CuraGen's Golden (in 2002) was in no doubt that deciphering the druggable genome sequence will ultimately restock the therapeutic pipeline.

It is clear from the distribution of the gene-family populations that there are no undiscovered large protein families, which indicates that remaining targets will be members of very small families. Clearly, the number of potential protein targets could be larger than the number of genes, owing to post-translational modifications and the assembly of functional complexes; however, this is not likely to increase the number of specific drug-binding sites.

Furthermore, the question then becomes which specific target will be *druggable and disease-modifying*. It is estimated that about 10% of the entire human genome is involved in disease onset or progression, resulting only in approximately 3000 potential targets suitable for therapeutic intervention. The key is to discriminate between the disease-associated versus disease-modifying genes. Hajduk et al. [9] suggested a way to employ Systems Biology to identify potential disease targets.

All drugs somehow interfere with signal transduction, receptor signaling and biochemical equilibrium. Additionally, for many drugs we know, and for most we suspect, that they interact with more than one target (pleiotropy). So, there will be simultaneous changes in several biochemical signals, and there will be feedback reactions from the perturbed pathways. In most cases, the *net result will not be linearly deductible from single effects and SB tools must be applied*. In addition, this phenomenon is a source of off-target effects.

For drug combinations, this is even more complicated. Thus, a clinically relevant 'target' might consist not of a single biochemical entity, but of simultaneous interference of a number of receptors (pathways, enzymes and so on). And, only the net clinical effect will be beneficial. Such target definition, not derived from a single direct chemical interaction *will require input from systems biology* [10]. Clinical input is crucial: the ability of a protein to bind a small molecule with the appropriate chemical properties at the required binding affinity might make it druggable, but does not necessarily make it a potential drug target, for that honor belongs only to proteins that are also linked to disease.

The results of screening in the pharmaceutical industry and the limited number of druggable targets suggest that, within the next decade, the industry could reach a position in which 'hits' or chemical leads are available for most potentially druggable targets. The challenge for the industry will then not necessarily be in the discovery of leads, but in *discovering and assessing the therapeutic utility of its leads and druggable targets (indications)* [11]. Furthermore, filtering of targets will become imperative as the space for potentially biologically relevant pharmacophores is enormous, and even large libraries populate it only sparsely, and attrition remains severe.

2.1.1 Bioinformatics Inputs (BI Inputs)

Drug-likeness and the Ro5 (see below) has been used to advantage, but as our knowledge of SB grows there is a need to move towards more predictive approaches. This will require prediction of where and how drugs interact with metabolism, which can be addressed by cheminformatics methods to assess molecular similarity between a putative drug and reactive metabolites. Concluding, the inputs of SB at identifying drug targets and BI are clearly needed in order to arrive at a target identification and validation.

2.2 Combinatorial Chemistry Tools

Combinatorial Chemistry methods facilitate:

- The synthesis of large libraries of small organic compounds and peptides
- The evaluation of these libraries against molecular and cellular targets
- The computer simulation of a large number of different but structurally related molecules

Combinatorial chemistry has contributed significantly towards the evolution of the DD process with the major emphasis in the field evolving towards methods to accelerate lead generation/lead optimization, focused parallel synthesis, high throughput technology, and chemical biology. It is justified on the assumption that the estimated size of the synthetically tractable space is in the order of 10^{20}–10^{24} molecules [12].

The latest report from a series of combinatorial chemistry surveys is available [13], providing bioactive libraries of different types. A new grouping highlights discovery and characterization of *molecular probes* or tool compounds allowing an *investigation of fundamental biological function at molecular, cellular and whole organism levels*. Such probes may be defined as small molecules that elicit a cellular event or phenotypical result through the specific interaction with a target protein, pathway or analyte. A good example is a breast cancer cell migration probe [14].

2.2.1 BI Inputs

Various computational tools are available as decision support tools for medicinal chemists involved in compound library synthesis programs [15]. These methods can be used to assemble a flexible library consisting of a structure-based library design followed by property-biased library refinement and final selection according to structure-activity-relationship considerations. Parallel Computing has facilitated the development of structure-based tools that are able to screen hundreds of thousands of molecules [16].

Further lead improvement is achieved with the help of diversity-oriented syntheses and asymmetric (chiral) compound preparation methods (see below).

2.2.2 Diversity Tools

Diversity can be accomplished chemically and biologically while seeking tools which can enrich the chemical space. Although purely chemically-based, diversity tools will have reached their full potential once the **SB and BI tools** are applied (as to any other OMICs) and explored.

2.2.2.1 Diversity-Oriented Synthesis and Discovery

Diversity-Oriented Synthesis (DOS) aims to prepare (mostly via solid-phase chemistry) collections of structurally complex and diverse compounds from simple starting materials, typically from 'split-pool' combinatorial chemistry. The DOS concept is typically applied through four elements of diversity:

- Building block
- Functional groups
- Stereochemistry, and
- Skeleton branching

Mimicking the broad structural features of natural products may allow the discovery of compounds that modulate the functions of macromolecules for which ligands are not known [17].

A simpler synthetic planning strategy is also available. In it a "retrosynthetic analysis" of the desired target structure is employed (which means a step-wise analysis of chemical transformations, starting from the complex target structure towards more simple starting materials) [18].

2.2.2.2 Differentially Expressed Proteins

DNA microarrays have allowed researchers a tool to detect differential expression of numerous, genes in a small sample and have clearly revolutionized the way gene expression analysis is now carried out (differentially expressed protein screen; detected against a normal background). These microarrays are, however, confined to the detection of gene transcripts and do not permit the analysis of the translational product (protein). The recent development of protein microarrays now offers the ability to simultaneously analyze the protein expression of several hundred proteins to measure the presence, biochemical characteristics and activation state of a considerable number of proteins in a single experiment.

2.2.2.3 Phage Display

Phage display (phage display screen) provides another diversity tool, based on the exploration of the most natural and efficient DD process. Molecular imaging is at the forefront in the advancement of in vivo diagnosis and monitoring of disease.

New peptide-based molecular probes to facilitate disease/cancer detection are rapidly evolving. Peptide-based molecular probes that target apoptosis, angiogenesis, cell signaling and cell adhesion events are in place. Phage display technology is commonly employed to identify peptides as tumor-targeting molecules. Numerous peptides that bind cancer cells and cancer-associated antigens have been reported from phage library selections.

Phage display screens have also been performed in live animals to obtain peptides with optimal stability and targeting properties in vivo. To this point, few in vitro, in situ, or in vivo selected peptides have shown success in the molecular imaging of cancer, the notable exception being vascular targeting peptides identified via in vivo selections. However, determining which peptide best translates into a useful drug is of particular importance. Examination of successfully marketed drugs has highlighted key features of a winning agent, including low molecular weight, high affinity, stability, solubility; lipophilicity and conformational rigidity (see Ro5).

A number of other platform strategies have been developed to screen polypeptide libraries for ligands targeting receptors selectively expressed in the context of various cell surface proteomes (cell display, ribosomal display, mRNA display and covalent DNA display) while phage display being by far the most utilized.

2.2.2.4 Chirality Tools

Advantages of using stereochemically pure drugs are that:

- Total dose could be reduced,
- The dose–response relationship would be simpler,
- A source of variability would be removed, and
- Toxicity from the inactive stereoisomer would be minimized.

Stereospecific interactions at recognition sites result in differences in both biological and toxicological effects. This fact underlies the continuing growth in chiral chemistry, rooted as it is in fundamental biochemistry.

The pharmaceutical industry has undergone a strategic shift and embraced the wide spectrum of asymmetrical synthetic methods, leading to chiral compounds, now available [19]. The use of these processes in developmental synthesis and large-scale manufacturing has provided new challenges in DD, motivated by a desire to improve industrial efficacy and decrease the time from the conception of a new drug to the market.

Chiral compounds are typically synthesized by asymmetrical synthesis [20], using a rapid screening method for monitoring specificity using both combinatorial chemistry and mass spectrometry.

BI Inputs: Few quantitative features are included in the chirality screen. Hologram quantitative structure-activity relationships (HQSAR) may be performed on a set of structurally diverse molecules with known human oral bioavailability.

The HQSAR model is useful for the design of new drug candidates with increased bioavailability as well as in the process of chemical library design, virtual screening, and HT screening [21].

2.2.2.5 Homology Tools at Discovery

Similarity metrics (homology modeling), and chemoinformatics are tools which further reduce the number of hits. They all employ some degree of quantitative data handling. Homology tools represent purely **BI input**.

Homology modeling (comparative modeling), refers to constructing an atomic-resolution model of the "target" protein and an experimental 3D structure of a related homologous protein (the "template"). To cite the author—"homology modeling relies on the identification of one or more known protein structures likely to resemble the structure of the query sequence, and on the production of an alignment that maps residues in the query sequence to residues in the template sequence. The sequence alignment and template structure are then used to produce a structural model of the target. Because protein structures are more conserved than DNA sequences, detectable levels of sequence similarity usually imply significant structural similarity" [22].

2.2.2.6 Cheminformatics

Cheminformatics combines chemistry and computer science (e.g., chemical graph theory) and mining the chemical space. Several selected tools are reviewed below:

- A data mining approach has been developed in order to construct a structure-diverse sub-library from the large PubChem (http://pubchem.ncbi.nlm.nih.gov/) database [23], suitable *in silico* virtual screening and in vitro HTS drug screening.
- Immunoinformatics, represents an emergent sub-discipline of BI, proposed to develop computational vaccinology as a potent tool in the quest for new vaccines.
- A text-based search engine allows efficient searching of compounds (in ChemDB; http://www.chemdb.com; containing nearly 5 million commercially available small molecules) based on over 65 million annotations from over 150 vendors [24].

2.2.2.7 Fragment Based DD

Fragment Based DD (FBDD) has developed over the last decade to take its place as a standard tool in the pharma industry [25]. It has yielded numerous, well-documented successes, and has proved to be the tool of choice for targets where much structural information is forthcoming, and which possess a well-defined, reasonably-small binding site [26]. To enable screening, a useful parameter has been introduced, the ligand efficiency, which is defined as the free energy of binding divided by the non-hydrogen atom count [27]. Further improvement follows, through so-called 'fragment evolution' and 'fragment linking'.

The next few years should see a maturing of the area, and as our understanding of how the concepts can be best applied increases. Undoubtedly, **BI input** will be introduced here in order to optimize the fragment chemistry.

2.2.3 Qualitative and Quantitative Screens and Filters

Several exclusionary filters have been devised to reduce the number of hits to manageable levels with the aim of identifying a lead substance. Rule of Five (Ro5) and extensions represent one such tool. Some screens have been devised that remove reactive chemical functionalities, on the premise that compounds having covalent chemistry capabilities have no place in DD (a simple qualitative rule or rule of thumb). The original Ro5 (Lipinski rule) deals with orally (lipidic) active compounds and defines the following simple physicochemical parameter ranges:

- $\log P \leq 5$
- H-bond donors ≤ 5
- H-bond acceptors ≤ 10,

These physicochemical properties are typically associated with 90% of orally active drugs that have achieved phase II clinical status. Essentially they are associated with acceptable aqueous solubility and intestinal permeability and comprise the first steps in oral bioavailability. This concept has been extended to include drug-like physicochemical properties [28]. Lead-like drugs refer to the screening of small MW libraries with detection of weak affinities in the high micromolar to millimolar range. A rule of three (Ro3) has been coined for these small molecule fragment screening libraries.

Simple kinetics (binding kinetics) in post Ro5 optimization/screen parameters in lead discovery and optimization is discussed below. Current lead discovery is mainly focused on identification of novel active and selective agonists, antagonists, or inhibitors that provide viable starting points into subsequent lead finding and optimization campaigns. The determination of binding kinetic profiles for efficacy and selectivity assessment in the context of lead discovery is widely under-appreciated, even though increasing evidence exists that renders *compounds that exhibit long binary complex residence time (in blood) better lead candidates* (Chap. 7). This behavior will highlight the underlying enzymological principles of so-called slow k (off) compounds (very low dissociation constant from the complex) and will provide examples that demonstrate that efficacious compounds display a distinct kinetic signature and will introduce a fragment-based lead discovery concept that guides medicinal chemistry towards long residence time kinase inhibitors for next-generation drugs.

2.2.3.1 BI Inputs

The use of computational rules as filters is an evolving field. A paradigm shift in drug discovery has resulted in the integration of Pharmacokinetics (PK) and DDv activities into the early stages of lead discovery [29]. In particular, *in silico* filters are used to help identify and screen out compounds that are unlikely to become drugs. Recent computational approaches towards the design of drug-like compound libraries, the prediction of drug-likeness, as well as intestinal absorption through passive transport, and the permeation of the blood-brain barrier serve as examples. The PK properties should also be included as important filters in virtual screening. Current development in theoretical models allows prediction of drug absorption-related properties, such as intestinal absorption, and blood-brain partitioning [30]. Both, chemical and computational methods serve as standard filter tools in the pharmaceutical industry.

2.2.4 Structure-Activity Methods (SAR/QSAR)

Quantitative structure-activity relationship (QSAR) is the process by which a chemical structure (a possible future pharmaceutical) is quantitatively correlated with a well defined process, such as biological activity or chemical reactivity (this effort has substantial **BI and SB inputs**). Biological activity can be expressed quantitatively as a concentration of a substance required to cause a certain biological response (this is just one aspect of the pure pharmacology, only half of the story; the key to characterization is equally functional impact). When physicochemical properties or structures are expressed by numbers, a mathematical relationship or quantitative structure-activity relationship, between the two can be established. The biological activity is usually measured in assays to determine the level of inhibition of particular signal transduction or metabolic pathways.

Unger and Hansch defined good practice in QSAR [31]:

- Select independent variables
- Justify the choice of the variables by statistical procedures,
- Apply the principle of parsimony,
- Have a large number of objects compared to the number of variables,
- Try to find a qualitative model of physicochemical or biochemical significance (chemical descriptors).

At present, more subtle quantitative methods are available:

- Chemical Computing Group Inc. developed novel 3D-QSAR descriptors allowing the 3D properties of compounds to be incorporated into traditional QSAR models, including Probabilistic Receptor Potentials to calculate the substrate's atomic preferences in the active site [32].
- Newer QSAR approaches [33]. The method seeks to characterize molecules by means of mathematical approaches used to establish a correlation between descriptor values and biological activity.
- New mathematical tools of Multivariate analysis include Artificial Neural Networks, Regression Trees, Random Forests, MARS (splines) and other methods (statistical). Various molecular descriptors have been used.
- The newest developments involve descriptors of 3D/4D QSAR, allowing for complex conformational methodology, covering a classification of drugs by their activity, MoA, target and binding site [34].

2.3 Summarizing

- This chapter discusses a mix of mostly reductionistic, qualitative and quantitative tools, at the chemistry discovery level.
- Chemical and other screening methods are a must because a typical compound entering a Phase I clinical trial undergoes years of rigorous preclinical testing but still only has about 8% change to reach the market.
- The number of druggable targets (number of specific drug-binding sites) is becoming exhausted and saturated by discovered drugs.
- The present insight into disease and biological processes is not sufficient to predict the therapeutic value of druggability of a novel target.
- While assessing the lead utility, the key is discrimination between disease-associated versus disease-modifying genes.
- Despite the initial promise of combinatorial chemistry, particularly large library combinatorial chemistry, to greatly accelerate drug discovery, this approach has not been fully effective as a means to build the compound collections of pharmaceutical and biotechnology companies.
- Combinatorial chemistry, though, has already had a great impact on DD programs and has contributed to the discovery of drugs that are currently in clinical trials or already in the market. Its significance will grow.
- Further lead improvement/screen can be achieved with the help of diversity-oriented syntheses and chiral compound preparation methods. Computer-based guidance will help.
- Several chemical BP tools are available to guide similarity searches.

- Application of qualitative rules such as rule-of-thumb (ROT) and computational filters will become more efficient. The huge number of active motifs can be reduced to a handful of promising lead series.
- QSAR is a well established tool where physicochemical and molecular descriptors are correlated with bioassay outputs.
- *In silico* tools will need to become a part of every pharmacologist's tool kit and this will require training in modeling and informatics, alongside in vivo, in vitro and molecular skills.
- A suite of QSAR models for making quantitative prediction of interactions of a test molecule with important classes of human proteins should be integrated into the Absorption, Distribution, Metabolism, Excretion, and Toxicity (ADMET) modeling effort (see Chap. 7).

References

1. Dixon SJ, Stockwell BR (2009) Identifying druggable disease-modifying gene products. Curr Opin Chem Biol 13(5–6):549–555
2. Borman S (2006) Chemical biology of the cell. Chem and Engg News 84(50):34–35
3. Paolini GV, Shapland RH, van Hoorn WP, Mason JS, Hopkins AL (2006) Global mapping of pharmacological space. Nat Biotechnol 24(7):805–815
4. Bakheet TM, Doig AJ (2009) Properties and identification of human protein drug targets. Bioinformatics 25(4):451–457
5. Musso GA, Zhang Z, Emili A (2007) Experimental and computational procedures for the assessment of protein complexes on a genome-wide scale. Chem Rev 107(8):3585–3600
6. Plewczynski D, Rychlewski L (2008) Meta-basic estimates the size of druggable human genome. J Mol Model 15(6):695–699
7. Drews J (1995) Intent and coincidence in pharmaceutical discovery: the impact of biotechnology. Arzneimittelforschung 45(8):934–939
8. Drews J (2003) Strategic trends in the drug industry. Drug Discov Today 8(9):411–420
9. Hajduk PJ, Huth JR, Tse C (2005) Predicting protein druggability. Drug Discov Today 10(23–24):1675–1682
10. Imming P, Sinning C, Meyer A (2006) Drugs, their targets and the nature and number of drug targets. Nat Rev Drug Discov 5(10):821–834 Erratum in Nat Rev Drug Discov 6(2):126 (2007)
11. Hopkins AL, Groom CR (2002) The druggable genome. Nat Rev Drug Discov 1(9):727–730
12. Ertl P (2003) Cheminformatics analysis of organic substituents: identification of the most common substituents, calculation of substituent properties, and automatic identification of drug-like bioisosteric groups. J Chem Inf Comput Sci 43(2):374–380
13. Dolle RE, Le Bourdonnec B, Goodman AJ, Morales GA, Thomas CJ, Zhang W (2007) Comprehensive survey of chemical libraries for drug discovery and chemical biology. J Comb Chem 10(6):753–802
14. Menzella HG, Reisinger SJ, Welch M, Kealey JT, Kennedy J, Reid R, Tran CQ, Metaferia BB, Chen L, Baker HL, Huang XY, Bewley CA (2007) Synthetic macrolides that inhibit breast cancer cell migration in vitro. J Am Chem Soc 129(9):2434–2435
15. Huwe CM (2006) Synthetic library design. Drug Discov Today 11(15–16):763–767
16. Schnur DM (2008) Recent trends in library design: 'rational design' revisited. Curr Opin Drug Discov Devel 11(3):375–380

17. Marcaurelle LA, Johannes CW (2008) Application of natural product-inspired diversity-oriented synthesis to drug discovery. Prog Drug Des 66:187–216
18. Kaiser M, Wetzel S, Kumar K, Waldmann H (2008) Biology-inspired synthesis of compound libraries. Cell Mol Life Sci 65(7–8):1186–1201
19. Burke D, Henderson DJ (2002) Chirality: a blueprint for the future. Br J Anaesth 88(4):563–576
20. Stevens SM Jr, Prokai-Tatrai K, Prokai L (2005) Screening of combinatorial libraries for substrate preference by mass spectrometry. Anal Chem 77(2):698–701
21. Moda TL, Montanari CA, Andricopulo AD (2007) Hologram QSAR model for the prediction of human oral bioavailability. Bioorg Med Chem 15(24):7738–7745
22. Schuffenhauer A, Floersheim P, Acklin P, Jacoby E (2003) Similarity metrics for ligands reflecting the similarity of the target proteins. J Chem Inf Comput Sci 43(2):391–405
23. Xie XQ, Chen JZ (2008) Data mining a small molecule drug screening representative subset from NIH PubChem. J Chem Inf Model 48(3):465–475
24. Chen JH, Linstead E, Swamidass SJ, Wang D, Baldi P (2007) ChemDB update full-text search and virtual chemical space. Bioinformatics 23(17):2348–2351
25. Leach AR, Hann MM, Burrows JN, Griffen EJ (2006) Fragment screening: an introduction. Mol Biosyst 2(9):430–446
26. Erlanson DA (2002) Fragment-based lead discovery: a chemical update. Curr Opin Biotechnol 17(6):643–652
27. Fattori D, Squarcia A, Bartoli S (2008) Fragment-based approach to drug lead discovery: overview and advances in various techniques. Drugs R D. 9(4):217–227
28. Lipinski C, Hopkins A (2004) Navigating chemical space for biology and medicine. Nature 432(7019):855–861
29. Matter H, Baringhaus KH, Naumann T, Klabunde T, Pirard B (2001) Computational approaches towards the rational design of drug-like compound libraries. Comb Chem High Throughput Screen 4(6):453–475
30. Hou T, Wang J, Zhang W, Wang W, Xu X (2006) Recent advances in computational prediction of drug absorption and permeability in drug discovery. Curr Med Chem 13(22):2653–2667
31. Unger SH, Hansch C (1973) On model building in structure-activity relationships A reexamination of adrenergic blocking activity of beta–halo–beta–arylalkylamines. J Med Chem 16(7):745–749
32. Esposito EX, Hopfinger AJ, Madura JD (2004) Methods for applying the quantitative structure-activity relationship paradigm. Methods Mol Biol 275:131–214
33. Tropsha A, Golbraikh A (2007) Predictive QSAR modeling workflow, model applicability domains, and virtual screening. Curr Pharm Des 13(34):3494–3504
34. Potemkin V, Grishina M (2008) Principles for 3D/4D QSAR classification of drugs. Drug Discov Today 13(21–22):952–959

Chapter 3
Integrative Systems Biology
I—Biochemistry: Phase I Lead Discovery
and Molecular Interactions

This chapter represents a mix of reductionist and holistic tools. Molecular screens and Biomimetics represent advanced reductionist approaches—the former are well established in the industry, although still developing. Similarly, the collateral efficacy/permissive antagonism concept may add to this effort, possibly generating new targets. Solving different co-drugging modalities represents a typical SB approach. Likewise, text mining does add to the holistic (global) effort. Tools to analyze biochemical networks and the phenomenon of emergence may lead to the establishment of 'new biology' or computational systems biology (CSB). Reactome analysis and bioinformatics tools only reinforce this effort. The level of development of the above quantitative tools is not uniform: some are advanced and mature (e.g., molecular screens), some require more inputs and are undergoing rapid evolution.

Definitions:

- **Molecular modeling** is a collective term that refers to theoretical methods and computational techniques to model or mimic the behavior of molecules.
- **Receptor** or **host** is the 'receiving' molecule, most commonly a protein, while **Ligand** or **guest** is the complementary partner which binds to the receptor.
- **Docking** is a computational simulation of a candidate ligand binding to a receptor, while **Scoring** is a process of evaluating a particular fit.
- **Biomimetic** compounds are synthetic materials with composition and properties similar to those made by living organisms.
- **Collateral efficacy** is a new concept, featuring differential effects depending on therapeutic niche.
- **Co-drugging** concept evolved from pleiotropy/redundancy effects on cell.
- **Text mining** refers to extracting knowledge from unstructured textual data.
- **Biochemical network** is an irreducible map of biochemical reactions. Similarly, other networks may be defined: transcriptional, proteomic, etc.

A. Prokop and S. Michelson, *Systems Biology in Biotech & Pharma*,
SpringerBriefs in Pharmaceutical Science & Drug Development,
DOI: 10.1007/978-94-007-2849-3_3, © The Author(s) 2012

- **Chemometrics** (e.g., Gemperline, 2006) involves, in very broad terms, classical and newer methods of statistical analysis, such as Multivariate analysis, principal components analysis (PCA), partial least-squares (PLS), followed by regression, clustering, and pattern recognition, lately, application areas have gone on to represent new domains, such as molecular modeling and QSAR, cheminformatics, the '-omics' fields of genomics, proteomics, metabonomics and metabolomics (some covered in later chapters). Typically, it is applied to solve both descriptive and predictive problems in experimental life sciences.

This section will discuss some theoretical, kinetic and quantitative concepts of biochemistry relevant to DD. First, we will cover reductionist phenomena of receptor-ligand interaction and molecular modeling, and then move onto biomimetic approaches at generating biologically active substances and onto collateral efficacy. SB will include the co-drugging concept, and text mining for interactions. In concluding, the chapter will review the current status of modeling of biochemical, signaling and interaction networks, a basis of SB. Wherever relevant, we will discuss quantitative tools, which are of great importance for building the fundamentals of SB.

3.1 Molecular Screens: Receptor–Ligand (R–L) Interaction and Molecular Modeling

This is an extension of the QSAR concept with the additional help of 3D pharmacophore modeling. With regards to small-ligand-receptor, in silico screening methods, one can usually distinguish two main strategies: ligand-based and receptor (structure)-based methods. This part exclusively uses a variety of **BI tools**.

The absence of a 3D receptor structure is the main reason for the application of ligand-based methods: similarity search; clustering; pharmacophore alignment (matching); 3D-pharmacophore modeling; and quantitative structure–activity relationship (QSAR). It uses information to find compounds that are known to bind to the desired target and then identifies other molecules in diverse databases with similar properties. Receptor-based methods essentially search for a ligand whose orientation and conformation achieves the highest degree of complementarity with respect to all details of the receptor steric constraints and interaction geometries (ligand docking) while the three-dimensional structure of the target is known either by X-ray crystallography or NMR experiments or predicted by homology.

Pharmaceutical research employs computer docking techniques for a variety of purposes, most notably in the virtual screening of large databases of available chemicals in order to select likely drug candidates: protein structure prediction, structural analysis, 'definition of ligand binding site' and *in silico* docking and scoring (e.g., Villoutreix et al. [1]. Several protein–ligand docking software packages are available (e.g., in [2, 3]), and the benefit of machine learning tools is that they are robust and model-free [4, 5].

Ligand binding to cell membrane receptors sets off a series of protein interactions that convey the nuances of ligand identity to the cell interior [6]. To study such interaction in living cells, different fluorescence imaging techniques are available (FRET, FCM and others). The challenge is extracting the quantitative information that is necessary to verify different models of signal transduction.

Complementary to experimental tools, we will cover detailed modeling of receptor-ligand interactions as an objective of molecular modeling tools below.

3.1.1 Molecular Modeling

Molecular modeling is often referred to as structural genomics and consists of determination of 3D structure of all proteins of a given organism, by experimental methods such as X-ray crystallography, NMR spectroscopy or computational approaches such as homology modeling. The simplest calculations can be performed by hand, but inevitably computers are required to perform molecular modeling on any reasonably sized system. The molecular modeling technique features the atomic level description of the molecular systems while reducing the complexity of the system. This is in contrast to quantum chemistry (also known as electronic structure calculations). The benefit of molecular modeling is that it reduces the complexity of the system.

3.1.2 Quantum Chemistry

Quantum chemistry mathematically describes the fundamental behavior of matter at the molecular scale, employing many approximations for most practical purposes. In quantum mechanics the Hamiltonian, or the physical state, of a particle can be expressed as the sum of two operators, one corresponding to kinetic energy and the other to potential energy. At present, quantum chemistry methods are prohibitively expensive.

3.1.3 Molecular Mechanics

Molecular mechanics (MM) methods are based on classical mechanics, allowing simulations to be performed on large systems containing more than 100 000 atoms. Free energy calculations derived from MM can account for flexibility for both the protein and the ligand as well as solvation effects (interaction with water), and accuracy and efficiency can be achieved within certain approximations. Huang et al. [7] list different methods/tools available.

3.1.4 Molecular Dynamics

Molecular dynamics (MD) simulations are now routinely applied to the study of biomolecular systems with the aim of sampling the configuration space more efficiently and getting a better understanding of the factors that determine structural stability and relevant biophysical and biochemical processes such as protein folding [8], ligand binding, and enzymatic reactions. This field has matured significantly in recent years.

3.1.5 Receptor Based QSAR Methods

Receptor based QSAR methods represent a computational marriage of structure activity relationship analysis and receptor structure based design that is providing valuable pharmacological insight to a wide range of therapeutic targets (Lushington et al. [9]). One implementation, called Comparative Binding Energy (COMBINE) analysis, is particularly powerful because of its explicit consideration of interatomic interactions between the ligand and receptor as the QSAR variable space. Other important methods account for covalent effects arising from ligand binding, as well as successful application of a COMBINE model to high throughput virtual screening. In many methods, lack of receptor flexibility considerations result in meaningless ligand binding scores, even when the correct receptor structure is obtained [10]. Modeling the role of the aqueous solvent in ligand–protein interactions is achieved via employment of one of three main computational techniques: free energy methods; ligand–protein docking and scoring; and the explicit inclusion of tightly bound water molecules in modeling (protonation) [11]. Using receptor conformations experimentally determined by crystallography or NMR or computationally, is a practical shortcut that may improve docking calculations. Binding site flexibility and protonation are issues which have been neglected for too long and require attention.

3.1.6 Biomimetics

Employment of natural compounds themselves should be the first *modus operandi* and the screening of natural products was in the forefront of early pharma efforts. Design of "mutein" proteins and biomimetic drugs (mimicking natural compounds) is beneficial as such substances can exhibit selective and sometimes novel biological properties. This is a purely chemical effort, not involving any modeling at this stage of development.

Demand for modified peptides with improved stability profiles and PK properties are driving extensive research effort in this field. Many structural modifications of peptides guided by rational design and molecular modeling have been established to develop novel synthetic approaches [12]. Tamerler and Sarikaya

[13] adapted combinatorial biology protocols to display peptide libraries, either on the cell surface or on phages, to select short peptides specific to a variety of practical materials systems. The protein epitope mimetic (PEM) approach (Robinson et al. [14]) considers folded 3D structures of peptides and proteins as a starting point for the design of synthetic molecules that mimic key epitopes involved in protein–protein and protein–nucleic acid interactions. PEM technology became a powerful tool for target validation.

3.2 Collateral Efficacy and Permissive Antagonism

The efficacy of a drug is generally determined by the drug's ability to promote a quantifiable biological response [15]. In contrast to classical receptor-occupancy theory, the concept that a single receptor can engage different signaling pathways and various drugs binding to this receptor might differentially influence each of these pathways led to the reassessment of the efficacy concept. Of particular note is the fact that ligands that behave as agonists toward a given signaling pathway can act, through the same receptor, as antagonists or even inverse agonists on a different pathway in the same cell. Specifically, some agonists might only partially activate the range of potential signaling systems in a cell or some antagonists might actively induce receptor internalization without activation. There is no longer justifiction for a linear view of efficacy whereby a single receptor activation state triggers all possible receptor interactions with a cell. Instead, a view of collateral efficacy, in which ligands can produce portions of the possible behaviors of receptors, is presented [16]. In practical terms, for example, a complex agonism is described, whereby a ligand produces positive agonism in quiescent systems and inverse agonism in constitutively active systems. Langmead [17] details a method for screening for positive allosteric modulators to examine a concentration–response (C/R) curve to the putative modulator in the presence of a single, low concentration of agonist.

There is a clear impact on DD, because the concept of permissive antagonism (a simple alternative allosteric model whereby the agonist and antagonist interact through conformational changes in the receptor protein) raises the possibility of selecting or designing novel ligands that differentially activate only a subset of functions of a single receptor, thereby optimizing therapeutic action. *Very rigorous kinetic and thermodynamic/equilibrium analysis* and **BI inputs will be required** to validate these allosteric effects.

3.3 Co-Drugging: Multiple Targets, Combination Therapy & Multistage Targeting

Advances in SB are revealing a phenotypic robustness and a network structure that strongly suggests that partial inhibition of a surprisingly small number of targets can be more efficient than the complete inhibition of a single target. This and the *success stories of multi-target drugs and combinatorial therapies suggest that systematic drug-design strategies should be directed against multiple targets.* The final effect of multiple drug action might often surpass that of complete drug action at a single target.

3.3.1 Multicomponent Drugs

Multicomponent drugs are standard in cytotoxic chemotherapy, but their development has required extensive empirical testing. However, experimentally validated numerical models should greatly aid the formulation of new combination therapies, particularly those tailored to the needs of specific patients [18]. Mathematical analysis is potentially powerful because many drug combinations can be explored computationally at much lower cost than in preclinical or clinical experiments. Some success has been achieved in formulating mathematical models of signaling pathways and oncogenic processes relevant to human disease. Such computational approaches to pharmacology require models that accurately recapitulate biochemical events in normal and abnormal (diseased) states. Some qualitative rules-of-thumb for selection were put forwards by Morphy et al. [19].

A clinical rationale for employing multiple ligands is based on the knowledge that several mutations are required for the development of cancers and other diseases, requiring several interventions. An enhanced (overlapping) toxicity of combination drugs might necessitate dose reduction of individual agents to ensure tolerability and reduced adverse effects. Clinical trials for combinations of novel targeted agents are listed in Dancey and Chen [20]. These authors suggested two different approaches:

- A single agent with multiple targets.
- A combination of agents with a single target (mixture of monotherapies).

3.3.2 Multi-Target Approach

Multi-target approaches suffer from the reductionist view as combination drugs are optimized for target selectivity and not for specific physiological responses. This is because our level of understanding of the effect of combination drugs is limited and it is not possible to predict the physiological consequences of modulating a novel (multiple) target. CSB offers a new solution, however. A real benefit will only come when a drug combination simultaneously impacts the principal and alternative target pathways of a disease. SB could provide excellent insights into these complex dynamics as the effects of multiple perturbations to any system are not obvious.

3.3.3 Multi-Stage Targeting

Multistage targeting is a concept that uses a sequence of drugs administered at specific doses and time intervals so that the dynamic state of target cells can be selectively perturbed into the system state that is desired for therapeutic purposes [21]. In such cases, it is important that the correct order and individual dose of the drugs to be administered, as well as the interval between the administrations of each drug, is examined. The multistage approach involves successively administering a set of drugs to differentiate between target and off-target cells. Good examples are sequential delivery of paclitaxel (Taxol; Bristol–Myers Squibb) and oxaliplatin (Eloxatin, Sanofi–Aventis), also termed as cancer chronotherapy. Another example is any compound that is cell cycle specific, e.g., acts to disrupt the molecular machinery in the S- or G2-phase of the cell cycle. Typically, these kinds of cycle specific compounds act optimally when the cell population has been "synchronized" to the target phase. In this case, Drug 1 inhibits progression through cycle at the boundary of the target phase and then, with appropriate timing, depending on the PK/PD of the Drug 1, Drug 2 can exert its maximum influence on the "resynchronized" cell populaiton. This strategy has been employed in cancer chemotherapy since the early works by Trucco and Rubinow ([22, 23]; see also Leith et al. [24]). Specifically, Mitomycin C in heterogeneous tumor growth acts in a similar way [24].

The assumption behind this approach is that there might be differences between target and off-target cells regarding their response to drugs, and that successive intervention, rather than a single intervention, could better exploit such differences to attain a greater level of selectivity. The next important step is the introduction of quantitative approaches (**BI and SB inputs**).

3.4 Text Mining for Interactions

Text mining for interactions is an emerging field concerned with the process of discovering and extracting knowledge from unstructured textual data, contrasting it with data mining which discovers knowledge from structured data. Text mining comprises three major activities: information retrieval, to gather relevant texts; information extraction, to identify and extract a range of specific types of information from texts of interest; and data mining, to find associations among pieces of information extracted from many different texts. Text mining aids in the construction of hypotheses from associations derived from vast amounts of text that are then subjected to experimental validation by experts (Sarić et al. [25]). Text mining relies entirely on **BI inputs**. Massive extraction databases are available: e.g., WikiProfessional http://www.wikiprofessional.org. Open questions remain on ontologies, database curation, on document processing and structure, and evaluation [26].

3.5 Employment of Biochemical Networks

Metabolism is based on elementary biochemical reactions, and forms the irreducible elements of a dynamic and adaptive network. The simulation of metabolic processes is based on specific models, which can be subdivided into abstract, discrete, and analytical. The abstract models typically employ automata and logical models, which permit the global discussion of fundamental aspects. The goal of analytical models is exact quantitative simulation, where the analysis of kinetic features of enzymes is important.

For example, glycolysis has been modeled by a set of differential equations. Discrete models employ state transition diagrams. Simple models of this class are based on simple production units, which can typically be combined. The graphical model allows the discussion of metabolic regulation processes and is representative for the class of graph theoretical approaches.

The above approach can be extended to allow for modeling of dynamic processes. In this case Petri Nets are used. Reddy et al. [27] presented the very first application of Petri Nets in molecular biology; this formalization is also able to model metabolic pathways. The highest abstraction level of this model class is represented by expert systems (Brutlag et al. [28]) and object oriented systems [29]. Expert systems and object oriented systems are developed by higher programming languages (Lisp, C++) and allow the modeling of metabolic processes by facts/classes (proteins and enzymes) and rules/classes (chemical reactions), by using a grammatical formalization that is able to model complex metabolic networks [30].

3.6 Overview of Deterministic Models

A quick overview of quantitative biochemical models (deterministic) is presented below. A full reaction mechanism is the set of elementary steps that specify how a reaction takes place. Elementary steps are those that cannot be decomposed into a more detailed scheme. In fact, a biochemical system is a network of elementary steps connecting various reactants, intermediates and products. A comprehensive description of such networks determines the number of chemical species and processes, the sequence of interactions and the rate laws governing the elementary reaction velocities.

Mass action law and linearized form are key concepts. It is now common to employ linearized forms of classical Michaelis–Menten enzyme kinetic equations to tackle metabolic and signaling pathways (Crampin et al. [31]). The assumption is that most reactions proceed at a high rate and no significant accumulation of intermediates occurs. Exceptionally, modelers have used non-linear descriptions for branching points at signaling or when crossing boundaries (transport), or when concentration-dependent clearance and binding/dissociation (equilibrium) are considered. A linear system is easier to simulate because of easier parameter identification and process simulation [32]. Additionally, linear systems allow for a clearer and more in-depth analysis of system steady states and their stability characteristics.

Assessing the bidirectionality of enzyme reactions and equilibrium is another kinetic aspect which must be described. The major problem of biology, as well as the 'inverse problem' of determining parametric causes from measured effects (variables), to which it is related, is understanding, at a lower level, the time-dependent changes of state that are commonly described at a higher level of organization; an issue often referred to using terms such as 'self-organization', 'emergence', networks and complexity [33].

3.6.1 Emergence

Ricard [34] states that a complex state may, because of some novel unpredictable properties, *emerge as a consequence of the interaction of the components (parts) of the system*. For biochemical systems, a system composed of two components displays negative or positive integration and is defined as a *complex system*. Such a system structure implies it cannot be reduced to its components.

Several recently constructed detailed kinetic models of metabolism (glycolysis), signal transduction (EGF receptor), and the eukaryotic cell cycle (*Saccharomyces cerevisiae*) have been used to exemplify the Silicon Cell project. These models are stored and made accessible via the JWS Online Cellular Systems Modeling, a web-based repository of kinetic models, together with a user-friendly graphical interface. The goal is to combine models on parts of cellular systems and ultimately to construct detailed kinetic models at the cellular level [35], at this moment limited to microbial systems. There is a need to develop and provide access to mammalian systems models (note some signaling pathways have been studied in some detail, see Oda et al. [36] on EGFR pathway maps). Curating repository models is another challenge. This type of approach we call 'reaction network' analysis (see also below in Chap. 5).

Although ODEs are commonly used to model signaling systems such as the ERK, they have one major drawback, and that is they are reliant on high-frequency sampling and absolute parameter data, such as kinetic rates and absolute initial concentrations. However, *a lot of the data generated by biologists, including data generated from HT techniques, are not directly amenable to modeling*, as they often contain sparse time series, are qualitative rather than quantitative, and show relative changes rather than changes in absolute concentrations.

Two additional (steady state) approaches are FBA (Flux Balance Analysis—Bonarius et al. [37]) and MCA (Metabolic Control Analysis [38]). FBA is an approach to constrain a metabolic network based on the stoichiometry of metabolic reactions and does not require kinetic information. MCA is a quantitative analysis of fluxes and concentrations, The relative control exerted by each step on a system variable is measured by applying a perturbation to the step and measuring the effect on the variable of interest after the system has settled to a new steady state. An extension of MCA has been recently suggested for dynamics of signal transduction (Hornberg et al. [39]). Thermodynamically-constrained metabolic flux analysis for network study has been proposed [40]. Other constraints include, typically, reversibility, charge balance, redox balance, regulatory rules or volume constraints (of the cell).

Some qualitative approaches are also available. Hybrid intelligent or rule-based models (fuzzy logic, neural nets, genetic algorithms and statistical analysis) can easily span several scales (time and length) and be used as a tool for aiding human reasoning when many interacting variables participate in complex system [41]. Such approach enables the incorporation of (qualitative) biological expertise into the modeling process. The employment of qualitative methods is warranted because of tremendous difficulties in handling multiscale systems mathematically.

3.6.2 Reactome

The following paragraph represents a rather short survey of some database resources one might use for his/her research. Reactome is an online bioinformatics database of human biology described in molecular terms, covering DNA replication, transcription, translation, the cell cycle, metabolism, and signaling cascades and can be browsed (in the public domain) to retrieve up-to-date information about a topic of interest (http://www.reactome.org) and is periodically updated (Matthews et al. [42]). However, it does not include metabolome data. KEGG (Kyoto Encyclopedia of Genes and Genomes http://www.genome.jp/kegg/ is hypertext-based information on biochemical pathways, including metabolic and regulatory pathways, cell cycle and growth factor signaling. It is *not*, however, *queryable*. BioMart (http://www.biomart.org) is a query-oriented data management system developed jointly by the Ontario Institute for Cancer Research (OiCR) and the European Bioinformatics Institute (EBI). The system can be used with any type of data and is particularly suited for providing 'data mining' like searches of complex descriptive data. The Online Bioinformatics Resources Collection (OBRC) contains annotations and links for 2348 bioinformatics databases and software tools (www.hsls.pitt.edu/guides/genetics/obrc).

3.7 Bioinformatics

Bioinformatics (BI) is a relatively new discipline dealing with the computational needs of biology, which has become a highly data-intensive activity. Biology databases must deal with both variety and scale, as well as be able to integrate the disparate databases that are their information sources. At the same time they must

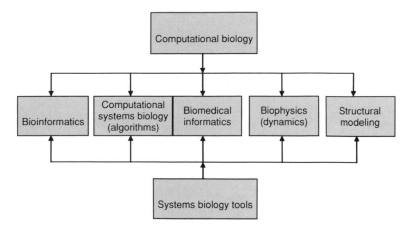

Fig. 3.1 Computational biology and its branches

provide flexible, friendly user interfaces for querying and data mining and cope with incomplete and uncertain data [43]. BI tools are very different from SB. However, both BI and SB, though somewhat different are complementary. Specifically, various BI computational methods address a broad spectrum of problems in functional genomics and cell physiology, including; analysis of sequences (alignment, homology discovery, gene annotation), gene clustering, *pattern recognition/discovery in large-scale expression data, elucidation of genetic regulatory circuits, analysis of metabolic networks and signal transduction pathways, which may in some instances* overlap with SB's goals.

In our view, bioinformatics is a part of larger field, computational biology:

- BI (incl. computational genomics).
- Structural modeling: molecular modeling and protein structure (some people consider structural tools as part of BI).
- Biophysics: molecular dynamics (Physical Biology of the Cell).
- Computational modeling of systems: CSB.

An overview of branches of computational biology is presented in Fig. 3.1. The question is: where is there an overlap between SB and BI? And though the disciplines exist in parallel, some of the tools are used interchangeably. Biomedical informatics (BMI) is another branch of CB, like biophysics.

Biophysics focuses on physical concepts and phenomena that cut across multiple biological structures and functions (Phillips et al. [44]), and three distinct approaches have been identified:

1. Mechanical, chemical equilibrium, entropy, statistics and tensegrity tools for resting cells;
2. Statistical, chemical rate, and electrochemical tools for cell dynamics; and
3. Networking in space and time.

The SB inputs (via the systems approach) are evident in all disciplines. Although the distinction is used by NIH in their working definitions of Bioinformatics and Computational Biology, it is clear that there is a tight coupling of developments and knowledge between the more hypothesis-driven research in computational biology and technique-driven research in bioinformatics.

3.8 Summarizing

- Molecular screens based on receptor-ligand interaction and molecular modeling represents an extension of the QSAR concept (with or w/o 3D structure).
- Binding site flexibility and protonation are issues which have been neglected for too long.
- Molecular dynamics tools will be soon integrated into QSAR methods.
- Biomimetic drug design is a purely chemical effort, not involving any modeling at this stage of development.
- Selective compounds, compared with multitarget drugs, may exhibit lower than desired clinical efficacy, as network analysis shows (see Chaps. 4 and 5).
- Real benefit will come only when a drug combination simultaneously impacts the principal and alternative targets of a disease.
- Collateral efficacy and permissive antagonism concepts may provide new DD leads once verified in vivo and established as a kinetic tool.
- Text mining aids in the construction of hypotheses from associations derived from vast amounts of text that are then subjected to experimental validation.
- Further effort is warranted to unify (and curate) repository network models.
- Many computational tools exist for visually exploring biological networks (35 existing tools reviewed by [45]).

References

1. Villoutreix BO, Renault N, Lagorce D, Sperandio O, Montes M, Miteva MA (2007) Free resources to assist structure-based virtual ligand screening experiments. Curr Protein Pept Sci 8(4):381–411
2. Seifert MH, Lang M (2008) Essential factors for successful virtual screening. Mini Rev Med Chem 8(1):63–72
3. Stoermer MJ (2006) Current status of virtual screening as analysed by target class. Med Chem 2(1):89–112
4. Zheng XF, Chan TF (2002) Chemical genomics: a systematic approach in biological research and drug discovery. Curr Issues Mol Biol 4(2):33–43
5. Fox T, Kriegl JM (2006) Machine learning techniques for in silico modeling of drug metabolism. Curr Top Med Chem 6(15):1579–1591
6. Zal T (2008) Visualization of protein interactions in living cells. Adv Exp Med Biol 640:183–197

7. Huang N, Kalyanaraman C, Bernacki K, Jacobson MP (2006) Molecular mechanics methods for predicting protein-ligand binding. Phys Chem Chem Phys 8(44):5166–5177
8. Marco E, Gago F (2007) Overcoming the inadequacies or limitations of experimental structures as drug targets by using computational modeling tools and molecular dynamics simulations. Chem Med Chem 2(10):1388–1401
9. Lushington GH, Guo JX, Wang JL (2007) Whither combine? New opportunities for receptor-based QSAR. Curr Med Chem 14(17):1863–1877
10. Totrov M, Abagyan R (2008) Flexible ligand docking to multiple receptor conformations: a practical alternative. Curr Opin Struct Biol 18(2):178–184
11. Mancera RL (2007) Molecular modeling of hydration in drug design. Curr Opin Drug Discov Devel 10(3):275–280
12. Vagner J, Qu H, Hruby VJ (2008) Peptidomimetics, a synthetic tool of drug discovery. Curr Opin Chem Biol 12(3):292–296
13. Tamerler C, Sarikaya M (2007) Molecular biomimetics: utilizing nature's molecular ways in practical engineering. Acta Biomater 3(3):289–299
14. Robinson JA, Demarco S, Gombert F, Moehle K, Obrecht D (2008) The design, structures and therapeutic potential of protein epitope mimetics. Drug Discov Today 13(21–22):944–951
15. Galandrin S, Oligny-Longpré G, Bouvier M (2007) The evasive nature of drug efficacy: implications for drug discovery. Trends Pharmacol Sci 28(8):423–430
16. Kenakin T (2007) Allosteric agonist modulators. J Recept Signal Transduct Res 27(4):247–259
17. Langmead CJ (2007) Screening for positive allosteric modulators: assessment of modulator: assessment of modulator concentration-response curves as a screening paradigm. J Biomol Screen 12(5):668–676
18. Fitzgerald JB, Schoeberl B, Nielsen UB, Sorger PK (2006) Systems biology and combination therapy in the quest for clinical efficacy. Nat Chem Biol 2(9):458–466
19. Morphy R, Kay C, Rankovic Z (2004) From magic bullets to designed multiple ligands. Drug Discov Today 9(15):641–651
20. Dancey JE, Chen HX (2006) Strategies for optimizing combinations of molecularly targeted anticancer agents. Nat Rev Drug Discov 5(8):649-59.
21. Kitano H (2007) A robustness-based approach to systems-oriented drug design. Nat Rev Drug Discov 6(3):202–210
22. Truco E (1965) Mathematical models for cellular systems: the Von Foerster equation. Part I Bull Math Biophys 27:283–303
23. Rubinow SI (1968) A Maturity-Time Representation for Cell Populations. Biophys J 8:1055–1073
24. Leith JT, Faulkner LE, Bliven SF, Michelson S (1988) Compositional stability of artificial heterogeneous tumors in vivo: use of mitomycin C as a cytotoxic probe. Cancer Res 48(10):2669–2673
25. Sarić J, Engelken H, Reyle U (2008) Discovering biomedical knowledge from the literature. Methods Mol Biol 484:415–433
26. Altman RB, Bergman CM, Blake J, Blaschke C, Cohen A, Gannon F, Grivell L, Hahn U, Hersh W, Hirschman L, Jensen LJ, Krallinger M, Mons B, O'Donoghue SI, Peitsch MC, Rebholz-Schuhmann D, Shatkay H, Valencia A (2008) Text mining for biology–the way forward: opinions from leading scientists. Genome Biol 9(Suppl 2):S7
27. Reddy VN, Mavrovouniotis ML, Liebman MN (1993) Petri net representation in metabolic pathways. In: Hunter L et al (eds) Proceedings first international conference on intelligent systems for molecular biology. AAAI Press, Menlo Park, pp 328–336
28. Brutlag DL (1989) Expert system simulations as active learning environments. In: Colwell RR (ed) Biomolecular data: a resource in transition. First CODATA Workshop on nucleic acid and protein sequencing data, Gaithersburg, MD, 1987, 367pp. Oxford University Press, New York, Oxford, pp 185–188

29. Stoffers HJ, Sonnhammer EL, Blommestijn GJ, Raat NJ, Westerhoff HV (1992) METASIM: object-oriented modelling of cell regulation. Comput Appl Biosci 8(5):443–449
30. Collado-Vides J (1991) A syntactic representation of units of genetic information—a syntax of units of genetic information. J Theor Biol 148:401–429
31. Crampin EJ, Schnell S, McSharry PE (2004) Mathematical and computational techniques to reduce complex biochemical reaction mechanisms. Progr Biophys Molec Biol 86:77–112
32. Kholodenko BN (2006) Cell-signalling dynamics in time and space. Nat Rev Mol Cell Biol 7(3):165–176
33. Bradley P, Misura KM, Baker D (2005) Toward high-resolution de novo structure prediction for small proteins. Science 309:1868–1871
34. Ricard J (2004) Reduction, integration and emergence in biochemical networks. Biol Cell 96(9):719–725
35. Snoep JL (2005) The Silicon Cell initiative: working towards a detailed kinetic description at the cellular level. Curr Opin Biotechnol 16(3):336–343
36. Oda K, Matsuoka Y, Funahashi A, Kitano H (2005) A comprehensive pathway map of epidermal growth factor receptor signaling. Mol Syst Biol 2005(1):2005.0010. [Epub 25 May 2005]
37. Bonarius HPJ, Schmid G, Tramper J (1997) Flux analysis of underdetermined metabolic networks: the quest for the missing constraints. Trends Biotechnol 15:308–314
38. Kacser H, Burns JA (1973) The control of flux. Symp Soc Exp Biol 27:65–104
39. Hornberg JJ, Bruggeman FJ, Binder B, Geest CR, de Vaate AJ, Lankelma J, Heinrich R, Westerhoff HV (2005) Principles behind the multifarious control of signal transduction. ERK phosphorylation and kinase/phosphatase control. FEBS J 272:244–258
40. Kümmel A, Panke S, Heinemann M (2006) Systematic assignment of thermodynamic constraints in metabolic network models. BMC Bioinf 7:512
41. Bosl WJ (2007) Systems biology by the rules: hybrid intelligent systems for pathway modeling and discovery. BMC Syst Biol 15(1):13
42. Matthews L, Gopinath G, Gillespie M, Caudy M, Croft D, de Bono B, Garapati P, Hemish J, Hermjakob H, Jassal B, Kanapin A, Lewis S, Mahajan S, May B, Schmidt E, Vastrik I, Wu G, Birney E, Stein L, D'Eustachio P (2009) Reactome knowledgebase of human biological pathways and processes. Nucl Acid Res 37(Database issue):D619–22
43. van Gend C, Snoep JL (2008) Systems biology model databases and resources. Essays Biochem 45:223–236
44. Phillips KA, Van Bebber S, Issa AM (2006) Diagnostics and biomaker development: priming the pipeline. Nat Rev Drug Discov 5(6):463–469
45. Suderman M, Hallett M (2007) Tools for visually exploring biological networks. Bioinformatics 23(20):2651–2659

Chapter 4
Integrative Systems Biology
II—Molecular Biology: Phase 2 Lead
Discovery and *In Silico* Screening

Many different OMICs/HTS techniques now allow huge amounts of molecular signatures to be collected and then analysed further by system tools. Among them, ChIP-on-chip is used to investigate interactions between proteins and DNA in vivo [1]. Chemogenomics, morphogenics and synthetic biology are only in the early stages of development, but may contribute to target identification. A key SB tool, the reconstruction of biological networks, represents an emerging field, undergoing explosive expansion, and will likely enable efficient mapping of gene onto function.

Definitions:

- **High throughput screening (HTS)** is a method of high-productivity experimentation.
- **OMICs** mean a study of the totality of biology: e.g. genomics, proteomics, etc.
- **Metabolomics**: experimental study of global metabolite profiles in a system (cell, tissue, or organism) under a given set of conditions.
- **Chemogenomics**: the study of the interaction of functional biological systems with exogenous small molecules; the intersection of the biological and chemical spaces.
- **Library:** a collection of molecules in a stable form that represents some aspect of an organism. A common type of library is a DNA library or genomic library.
- **Morphogenics:** a platform process that employs a dominant negative MMR gene to create genetic diversity, a purely experimental tool.
- **Minimal cell/genome:** an experimental cell model having the minimum but sufficient number of components to be defined as living, while **Synthetic Biology** combines science and engineering to design novel biological functions.
- **Reconstructing biological networks** in the sub-cellular environment: network reconstruction via simulation allows for an in-depth insight into the molecular mechanisms of a particular organism while correlating genome with physiology.

A. Prokop and S. Michelson, *Systems Biology in Biotech & Pharma*,
SpringerBriefs in Pharmaceutical Science & Drug Development,
DOI: 10.1007/978-94-007-2849-3_4, © The Author(s) 2012

- **Reverse and forward genetics:** 'reverse' genetics starts with a purified target, then moves to the chemical library for binding activity and finally tests a molecule in vivo for physiological effects. Classical forward genetics starts with a mutation phenotype and works towards identifying the mutated gene.

4.1 OMICs

The ability to sequence whole genomes has taught us that our knowledge of gene function is rather limited with typically 30–40% of open reading frames having no known function. Thus, within the life sciences there is a need for determination of the biological function of these so-called orphan genes, some of which might possibly become molecular targets for therapeutic intervention. The search for specific mRNA, proteins, or metabolites, useful as diagnostic markers has also increased, as has the fact that these biomarkers may be useful in following and predicting disease progression or response to therapy.

At present, functional analyses include gene expression (transcriptomics), protein translation (proteomics), metabolite network dynamics (metabolomics) and more recently phenomics (to determine the phenotypic response of whole biological systems to environmental, pathophysiological and genetic perturbations). In addition, techniques for DNA methylation pattern, microRNA and siRNA knockdown OMICs are being developed.

HTS leverages developments in the areas of modern robotics, data analysis and control software, liquid handling devices, and sensitive detectors, allowing one to efficiently screen millions of compounds and to potentially identify tractable small molecule modulators of a given biological process or disease state and advance them into high quality leads [2]. HTS is commonly defined as automatic testing of potential drug candidates at a rate in excess of 200,000 compounds per week (uHTS). The aim is to test large compound collections (>500,000 compounds and as many as 2,000,000) for potentially active compounds ('hits') in order to allow further development of compounds for pre-clinical testing ('leads'). HTS technology has emerged over the last few years as an important tool for DD and lead optimization.

4.1.1 BI Inputs

HTS is not just an experimental platform, but is subjected to several BI tools to analyze data effectively. Computational advances in image analysis and technological advancements in general cell biology have extended the utility of HTS into target validation, including siRNA screening.

Transcriptomics, a genome-wide measurement of mRNA expression levels based on DNA microarray technology is one of the main fields of study. It can also be applied to the specific subset of transcripts present in a particular cell or the total set of transcripts in a given organism.

Proteomics can be used for characterizing alterations in protein abundance finding novel protein–protein and protein–peptide interactions, investigating formation of large macromolecular complexes, and elucidating temporal changes in protein composition and phosphorylation in signal transduction cascades. Further, quantitative proteomics can directly compare the activation of entire signaling networks in response to individual stimuli and discover critical differences in their circuits that account for alterations of cell response (in disease). This method now allows investigation of cellular mechanisms at the organismic level. At present the OMICs technologies still:

- Lack the proper dynamic range in quantitative assays,
- Lack the capacity to scale up to proper size of data and measurements, and
- Display inadequate specificity/sensitivity, failing to avoid inflated false negative and positive results.

4.1.2 Metabolomics

4.1.2.1 SB and BI Inputs

Metabolomics is the study of global metabolite profiles in a system (cell, tissue, or organism) under a given set of conditions. Metabolites are the result of the interaction of the system's genome with its environment and are not merely the end product of gene expression, but also form part of the regulatory system in an integrated manner.

4.1.2.2 Metabolic Profiling

SB and BI Inputs

Metabolic profiling (metabolomics/metabonomics) is the measurement in biological systems of the complement of low-molecular-weight metabolites and their intermediates that reflects the dynamic response to genetic modification and physiological, pathophysiological, and/or developmental stimuli. The measurement and interpretation of the endogenous metabolite profile from a biological sample (typically urine, serum, or biological tissue extract) have provided many opportunities to investigate the changes induced by external stimuli (e.g., drug treatment) or to enhance our knowledge of inherent biological variation within specific subpopulations. Such efforts are now being integrated into SB [3].

4.1.2.3 Peptidome

The low-molecular-weight range of the circulatory (blood) proteome is termed the 'peptidome', and may be a rich source of disease, and especially cancer-specific, diagnostic information because it is a 'recording' of the cellular and extracellular enzymatic events that take place at the level of the tissue/cancer-tissue microenvironment [4].

4.1.2.4 Proteome Analysis

"Inverse labeling" of proteome analysis, is a strategy is based on the principle of protein stable isotope labeling and mass spectrometric detection, a procedure to evaluate protein expression of a diseased or a drug-treated sample in comparison with a control sample. Two inverse labeling experiments are performed in parallel. The perturbed sample (by disease or by drug treatment) is labeled in one experiment, whereas the control is labeled in the second experiment, while compared differentially using mass spectrometry. This enables the detection of protein modifications responding to perturbation [5]. ^{18}O and ^{15}N labels are typically used.

BI Inputs

The PathoGenome Database facilitates DD by providing potential drug targets for new anti-infectives derived from high-quality DNA sequences, annotation, search tools and powerful bioinformatics platforms. It covers medically important microbial organisms. Internet access is established through the LabOnWeb.com search engine (http://pathogenome.net/index.php/Main_Page).

4.2 Chemogenomics

4.2.1 BI Inputs

Chemogenomics is the study of the interaction of functional biological systems with exogenous small molecules, or in a broader sense the study of the intersection of biological and chemical spaces [6]. The biological space is analyzed at various postgenomic levels (genomic, transcriptomic, proteomic or any phenotypic level). The goal of chemogenomics is the rapid identification of novel drugs and drug targets embracing multiple early phase DD technologies ranging from target identification and validation, through compound design and chemical synthesis to biological testing and ADME profiling [7]. By integrating all information available within a protein family (sequence, SAR data, and protein structure), chemogenomics can efficiently enable cross-SAR exploitation, directed compound selection and early identification of optimum selectivity panel members. Successful examples include two major protein families: protein kinases and G-protein-coupled receptors [8].

4.3 Morphogenics

The ability to modulate the DNA mismatch repair (MMR) processes (referred to as morphogenics) in model systems offers a powerful tool for generating functional diversity in cells and multicellular organisms via the perpetual genome-wide accumulation of randomized point and slippage mutation(s). Morphogenics is a platform process that employs a dominant negative MMR gene to create genetic diversity within defined cellular systems and results in a wide range of phenotypes, thus enabling the development and improvement of pharmaceutical products and the discovery of new pharmaceutical targets.

MMR is a highly conserved biological pathway that plays a key role in maintaining genomic stability [9]. Defects in MMR are associated with predispositions to certain types of cancer including hereditary non-polyposis colorectal cancer, resistance to certain agents, and abnormalities in meiosis and sterility in mammalian systems [10].

The maintenance of genomic stability is the prime requirement for the stem cell phenotype and its normal functioning [11]. The increased mutation rate and absence of MMR may give rise to stem cell failure, a proliferative advantage and cancer stem cell formation. The importance of MMR in designing the therapeutic strategies specifically targeting tumor stem cells is being explored and pursued in clinical trials (e.g., in chronic myeloid leukaemia) [12].

The molecular mechanisms of the DNA MMR system have been uncovered over the last decade, especially in prokaryotes [13]. This is a rapidly developing field. The selective manipulation of the MMR process is a platform technology that offers many advantages for the discovery of druggable targets, as well as for the development of novel pharmaceutical products [14].

4.4 Minimal Phenotype and Synthetic Biology

Projects aiming at simplifying living cells converge with efforts to make synthetic genomes for minimal cells. Here, the aim is to synthesize a cell model having the minimal but sufficient number of components to be defined as living. One of the key aims of synthetic biology is to design and build a viable cell with a minimized genome that can serve as a platform for the construction of microbial factory cells for biotechnological applications.

4.4.1 SB and BI Inputs

Comparative genomics, using computational and experimental methods, enables the identification of the minimal set of genes that is necessary and sufficient for sustaining a functional cell [15]. For most essential cellular functions, two or more unrelated or distantly related proteins have evolved; only about 60 proteins, primarily those involved in translation, are common to all cellular life.

For example, a codon-optimized synthetic gene cluster of polyketide synthase genes has been synthesized to enable combinatorial makrolide (a group of anti-biotics) biosynthesis [16–18]. Such an approach offers a synthetic route for production of pharmaceutical intermediates. Cell-free synthetic biology, via discovery by design is now available.

Modeling can facilitate the design of engineered biological systems by allowing synthetic biologists to better predict system behavior prior to fabrication of self-replicating biosystems, such as in vitro genome replication, in vitro transcription, RNA processing and RNA modification, and, post translation, integration into a minimal cell for development of new drugs [19]. Synthetic biological engineering is emerging from biology as a distinct discipline based on quantification [20, 21], with enhanced emphasis on system behavior. This field is only just emerging and is very new.

4.5 Reconstructing Biological Networks

Both BI and SB contribute to network reconstruction, which is defined as a process of integrating different data sources into a unified representation of the chemical (and physical and mechanical) events that underlie a biochemical network [22].

Biochemical system (network) reconstruction (BSR) is emerging as an important tool as systems analysis of metabolic and growth functions in microbial organisms is rapidly developing and maturing. Such studies are enabled by reconstruction, at the genomic scale, of the biochemical reaction networks that underlie cellular processes. The network reconstruction process is organism specific and is based on an annotated genome sequence, high-throughput network-wide data sets and bibliomic data on the detailed properties of individual network components, as noted by the authors [23]. BSR consists of automatic genome annotation and automatic reconstruction. In parallel, high-throughput expression data paired with computational algorithms can be used to infer the structure of network interactions and the existence of causal relationships capable of reproducing the observed experimental data [24]. Additionally, graph-theoretical analysis of these networks, and the extension of static networks into various dynamic models capable of providing a new layer of insight into the functioning of cellular systems is now available [25, 26]. Qualitative and continuous or hybrid models, as well as stochastic Petri Nets models, have also been applied [27, 28] to these networks. Sackmann et al. defined a signal transduction pathway based on the network structure only. For this purpose, they introduced the new notion of feasible t-invariants, which represents the minimal self-contained subnets that are active under a given input situation. Each of these subnets stands for a signal flow in the system and represents a net decomposition into smallest biologically meaningful functional units.

A flux balance analysis (FBA)-based strategy, requiring an integrated stoichiometric reconstruction of signaling, metabolic, and regulatory processes [29], and referred to as integrated dynamic FBA (idFBA) simulates cellular phenotypes arising from integrated networks. Network reconstruction can be approached in three different ways, the last being the only quantitative route:

- Reconstruction of nodes, modules and pathways,
- Structural analysis via connectivity and casual reconstruction by means of linear pathways that connect signaling input to signaling output [30, 31], and
- Stoichiometric reconstruction [32].

In their 2007 paper, Zhang et al. [33] discuss the advent of generalized reconstruction methodology. They suggest that the gene network reconstruction can be improved by combining ontology with a clustering algorithm based on similarity. Mapping function to gene products in the genome consists of two steps: *ontology building* and *ontology annotation*. Ontology building is the formal representation of a domain of knowledge; ontology annotation is the association of specific genomic regions (including regulatory elements and products such as proteins and functional RNAs) to parts of the ontology. Typically, two complementary representations of gene function are employed: the Gene Ontology (GO) and the pathway ontology. Various theoretical constructs of mapping genes to functions, as derived from molecular observations, are depicted in Fig. 4.1. Note, crosstalk is a combination of pleiotropy and redundancy. Many phenotypic traits are properties that emerge from the collective action of several individual genes. Two of the nodes are perturbed (small red arrow) in this example. In case of cancer, a combination of two ligands (drugs) can shift the whole network into a more desired state, apoptosis and differentiation.

The employment of network analysis could be problematic for multifactorial diseases, as is the case of cancer [34]. If there is very little understanding of molecular components in the disease process and their dynamic associations, computation might be useless. However, when a fair amount of relevant information is available, computation can be quite good at assigning the pertinent functions. Another problem stems from the fact that the present GO annotations are poor for cancer genes. Because cancer can result from numerous multi-stage processes, trying to assign a function to an uncharacterized gene one at a time is hampered by the fact that the underlying correlation structure changes with time during cancer progression.

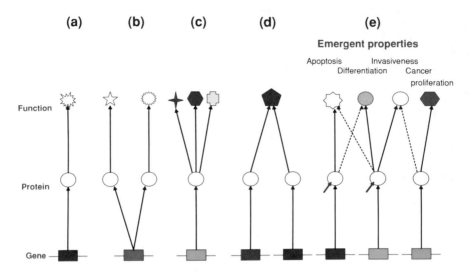

Fig. 4.1 Mapping of Gene to Function and Emergence. **a** Traditional linear genome-phenome relationship. **b** Divergent relationships may cause undesired side effects when acted upon in isolation. Alternative transcription may produce two distinct gene products or functions. Multi-component drugs can specifically inhibit two downstream pathways. **c** Pleiotropy when more than one gene product contributes to the same function. It can evoke unintended drug side effects, but might create an opportunity for multiple indications. **d** Redundancy (convergent relationships) when more than one gene product can compensate for inhibition of another pathway. Multi-component interventions must simultaneously inhibit both arms. **e** Crosstalk as an exhibit of emergence: the interaction between distinct signaling/regulatory pathways so that an input to one pathway has some effect on the output of the other

4.6 Summarizing

Different experimental techniques (OMICs), allow efficient collection of huge amounts of molecular signatures for each of the systems studied. HTS in target validation is now expanding its format to include data from many different OMICs environments, to better characterize targets and put them into a more biological context. Likewise, label-free technologies have recently attracted significant interest for sensitive and quantitative multiparameter analysis of biological systems. Chemical genomics is a new term that describes the development of target-specific chemical ligands and the use of such chemical ligands to globally study gene and protein functions. In the near future, identification of MMR (microsatellite instability-positive) profiles for tumors will enable targeting of MMR-deficient cells that are more tolerant to drugs. Interaction, biochemical, transcriptional and signaling network analysis may facilitate DD and guide delivery targets, while cellular pathway analysis is currently the prevalent method of SB. On the other hand, many signaling network models are still not available. The networks can be modeled by a variety of continuous, stochastic and discrete techniques, although differential equations are most common. Serious steps have been taken to provide a framework for quantitative characterization of kinetic, metabolic, regulatory and signaling networks. The next step, largely neglected, is to integrate these processes into a coherent qualitative and quantitative modeling tool(s) to devise a macroscopic (systems) description. However, huge gaps exist in terms of quantitative experimental characterization of networks in both normal and disease states, and the efforts are hampered by several key limitations in the accessible data [35]: small signal–noise ratios, insufficient time resolution, insufficient spatial resolution, and too few signals being measured. In the absence of rigorous data, initially, one should seek to *qualitatively reconstruct pathways* instead of aiming at developing fully detailed kinetic models; the task of detailed kinetic reconstruction based on time-series data alone is extremely difficult and often remains underdetermined (Chap. 5). Consequently, a large number of reconstructed systems are consistent with any given set of time-series data. To deal with this non-uniqueness, the solution space is often limited by a priori, reasonable assumptions such as linearity (in kinetic terms), sparseness (where sparseness refers to a representational scheme where only a few units—out of a large population—are effectively used to represent typical data vectors) or model structures are incorporated with predetermined constraints. A computational approach is available to deal with some experimental imperfections. Satish Kumar et al. [31] proposed systematic methods to identify and fill gaps in genome-scale metabolic reconstructions, by making modifications in the existing models and by adding missing reactions to allow for connectivity by comparing databases rich in reactions with existing genome-scale models. To enable discoveries, methods of rigorous experimental validation of computational tools finding need to be developed and introduced, particularly in terms of providing quantitative dynamic (perturbatory, time-course response profiles often referred to as metabonomics) measurements. By repeating the model building/learning and experimentation process, generation of more robust system descriptions will emerge as our understanding of the system (disease state) evolves.

References

1. Chen X, Jorgenson E, Cheung ST (2009) New tools for functional genomic analysis. Drug Discov Today 14(15–16):754–760
2. Ling XB (2008) High throughput screening informatics. Combinat Chem High Throughput Screen 11(3):249–257
3. Hollywood K, Brison DR, Goodacre R (2006) Metabolomics: current technologies and future trends. Proteomics 6(17):4716–4723
4. Petricoin EF, Belluco C, Araujo RP, Liotta LA (2006) The blood peptidome: a higher dimension of information content for cancer biomarker discovery. Nat Rev Cancer 6(12):961–967
5. Wang YK, Ma Z, Quinn DF, Fu EW (2001) Inverse 18O labeling mass spectrometry for the rapid identification of marker/target proteins. Anal Chem 73(15):3742–3750
6. Maréchal E (2008) Chemogenomics: a discipline at the crossroad of high throughput technologies, biomarker research, combinatorial chemistry, genomics, cheminformatics,

bioinformatics and artificial intelligence. Comb Chem High Throughput Screen 11(8):583–586

7. Birault V, Harris CJ, Le J, Lipkin M, Nerella R, Stevens A (2006) Bringing kinases into focus: efficient drug design through the use of chemogenomic toolkits. Curr Med Chem 13(15):1735–1748

8. Zheng XF, Chan TF (2002) Chemical genomics: a systematic approach in biological research and drug discovery. Curr Issues Mol Biol 4(2):33–43

9. Li LS, Morales JC, Hwang A, Wagner MW, Boothman DA (2008) DNA mismatch repair-dependent activation of c-Abl/p73alpha/GADD45alpha-mediated apoptosis. J Biol Chem 283(31):21394–21403

10. Nicolaides NC, Ebel W, Kline B, Chao Q, Routhier E, Sass PM, Grasso L (2005) Morphogenics as a tool for target discovery and drug development. Ann N Y Acad Sci 1059:86–96

11. Vaish M (2007) Mismatch repair deficiencies transforming stem cells into cancer stem cells and therapeutic implications. Mol Cancer 2(6):26

12. Larson SM, Godley LA (2010) Getting to the root of the stem cell in mutated chronic myeloid leukemia. Leuk Lymphoma 51(12):2147–2148

13. Jun SH, Kim TG, Ban C (2006) DNA mismatch repair system. Classical and fresh roles. FEBS J 273(8):1609–1619

14. Nicolaides NC, Ebel W, Kline B, Chao Q, Routhier E, Sass PM, Grasso L (2005) Morphogenics as a tool for target discovery and drug development. Ann N Y Acad Sci 1059:86–96

15. Werner E (2003) In silico multicellular systems biology and minimal genomes. Drug Discov Today 8(24):1121–1127

16. Santi DV (2006) Redesign, synthesis and functional expression of the 6-deoxyerythronolide B polyketide synthase gene cluster. J Ind Microbiol Biotechnol 33(1):22–28

17. Menzella HG, Reisinger SJ, Welch M, Kealey JT, Kennedy J, Reid R, Tran CQ, Metaferia BB, Chen L, Baker HL, Huang XY, Bewley CA (2007) Synthetic macrolides that inhibit breast cancer cell migration in vitro. J Am Chem Soc 129(9):2434–2435

18. Zotchev SB, Stepanchikova AV, Sergeyko AP, Sobolev BN, Filimonov DA, Poroikov VV (2006) Rational design of macrolides generated through in silico manipulation of polyketide synthases. J Med Chem 49(6):2077–2087

19. Forster AC, Church GM (2006) Towards synthesis of a minimal cell. Mol Syst Biol 2:45

20. Kaznessis YN (2007) Models for synthetic biology. BMC Syst Biol 1:47

21. Marchisio MA, Stelling J (2009) Computational design tools for synthetic biology. Curr Opin Biotechnol 20:479–485

22. Papin JA, Hunter T, Palsson BO, Subramaniam S (2005) Reconstruction of cellular signalling networks and analysis of their properties. Nat Rev Mol Cell Biol 6(2):99–111

23. Feist AM, Herrgård MJ, Thiele I, Reed JL, Palsson BØ (2009) Reconstruction of biochemical networks in microorganisms. Nat Rev Microbiol 7(2):129–143

24. Francke C, Siezen RJ, Teusink B (2005) Reconstructing the metabolic network of a bacterium from its genome. Trends Microbiol 13(11):550–558

25. Alon N, Dao P, Hajirasouliha I, Hormozdiari F, Sahinalp SC (2008) Biomolecular network motif counting and discovery by color coding. Bioinformatics 24(13):i241–i249

26. Christensen C, Thakar J, Albert R (2007) Systems-level insights into cellular regulation: inferring, analysing, and modelling intracellular networks. IET Syst Biol 1(2):61–77

27. Sackmann A, Heiner M, Koch I (2006) Application of Petri net based analysis techniques to signal transduction pathways. BMC Bioinformatics 2(7):482

28. Goss PJ, Peccoud J (1998) Quantitative modeling of stochastic systems in molecular biology by using stochastic Petri nets. Proc Natl Acad Sci U S A 95(12):6750–6755

29. Min Lee J, Gianchandani EP, Eddy JA, Papin JA (2006) Dynamic analysis of integrated signaling, metabolic, and regulatory networks. PLoS Comput Biol 4(5):e1000086

30. Nikiforova VJ, Willmitzer L (2007) Network visualization and network analysis. EXS 97:245–275

31. Satish Kumar V, Dasika MS, Maranas CD (2007) Optimization based automated curation of metabolic reconstructions. BMC Bioinformatics 20(8):212
32. Papin JA, Palsson BO (2004) Topological analysis of mass-balanced signaling networks: a framework to obtain network properties including crosstalk. J Theor Biol 227(2):283–297
33. Zhang S, Jin G, Zhang XS, Chen L (2007) Discovering functions and revealing mechanisms at molecular level from biological networks. Proteomics 7(16):2856–2869
34. Hu P, Bader G, Wigle DA, Emili A (2007) Computational prediction of cancer-gene function. Nat Rev Cancer 7(1):23–34
35. Rice JJ, Tu Y, Stolovitzky G (2005) Reconstructing biological networks using conditional correlation analysis. Bioinformatics 21(6):765–773

Chapter 5
Discovery: Computational Systems Biology (CSB) in Health and Disease I

To date, few cellular and gene networks have been reconstructed and analyzed in full. Examples include some prokaryotes and few eukaryotes for cellular networks. The methods currently used to analyze single database genomic sets are usually mature and refined. Network reconstruction is also enabled by analysing the molecular connectivity of a system by using correlation analysis. Additionally, monitoring the dynamics of the system and measuring the system's responses to perturbations such as drug administration or challenge tests can yield insights into the dynamics of the system. Microbial cells are fairly well characterized, but the status of similar efforts for mammalian cells is rather poor. While emergence can be conveniently studied via computational tools, the phenomenon of emergence is the single most important benefit of CSB.

Definitions:

- **Inferring network** (reverse engineering) is identical to cellular network reconstruction by means of computational and statistical methods.
- **Data mining** in informatics is the process of extracting hidden patterns from data into useful information to enable scientific discovery.
- **Gene regulatory network (GRN)** is a collection of DNA segments in a cell which interacts with others and with other substances, thereby controlling the rates at which genes in the network are transcribed into mRNA.
- **Cellular network** governs the basic processes of cell proliferation, differentiation and cell death.
- **Stem cells** are cells found in many multi-cellular organisms; they are characterized by the ability to self-renew through mitotic cell division and differentiation into a diverse range of specialized cell types.
- **Computational Systems Biology (CSB)** overlaps with SB, while applying computer science, applied maths and statistics to address biological problems.

A. Prokop and S. Michelson, *Systems Biology in Biotech & Pharma*,
SpringerBriefs in Pharmaceutical Science & Drug Development,
DOI: 10.1007/978-94-007-2849-3_5, © The Author(s) 2012

This chapter looks at how SB-enabled reconstruction of cellular networks could be inferred from dynamic biological experiments, and how BI in data mining can prioritize targets and expedite biological knowledge discovery. Then we describe the network clusters of diseases and discuss stem cell compartments as important targets for discovery and manipulation. The discussion then defines emergent properties in terms of interacting species, teaching us that it is network states that should be therapeutically targeted (defined on the basis of molecular and physiological networks and emergence) not individual genes or proteins. Finally CSB will be addressed and several important disease examples presented demonstrating the utility of the concept.

5.1 Cellular Environment: Network Reconstruction and Inference from Experimental Data

Biological networks are presented as nodes (proteins, genes and metabolites) in vector algebra connected by edges (protein–protein interaction, protein-gene, protein-compounds interactions and metabolic reactions). Depending on the type of data and interaction mechanisms, the edges are either directed or undirected. The network as a whole can be described as the average degree of edges, the average clustering coefficient, the shortest path between nodes and diameter (the longest distance). The analysis of yeast (and other) interactomes revealed that biological networks are remarkably non-random and the distribution of edges is very heterogeneous, with few highly connected nodes (hubs with 5–25 nodes) and the majority of nodes having very few edges. Such topology is defined as *scale-free*. The hubs are predominantly connected to low-degree nodes, a feature that gives the biological networks the property of robustness. A removal of substantial fractions of nodes still leaves the network reasonably intact. Few highly connected proteins in hubs, typically, contain a fairly high frequency mutated genes (particularly in the dysregulated network in diseased cells such as cancer) and play a central role in mediating interactions among numerous, less connected proteins (the less frequently altered ones). In addition, it is now recognized that these *hubs* (protein or gene hubs) are essential to biological function, preferentially encoding disease states and might be new prognostic or diagnostic biomarkers, and/or candidate *therapeutic targets* for intervention and targeting [1, 2].

The key property of biological networks is their *modular* nature, meaning that various kinds of cellular functionality are provided by relatively small tightly connected subnetworks of molecules. Network analysis and identification of the component modules is the key to network computational restructuring. Measures to determine overall network characteristics include the degree of clustering and the clustering coefficient, which identify networks that possess random, scale-free, or hierarchical structures. The latter two are frequent in biological systems.

Fig. 5.1 Star, tree, ring and connected topologies

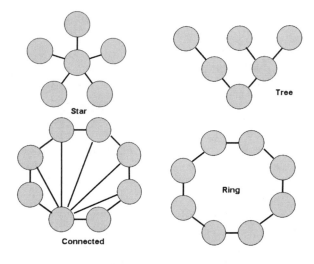

Hierarchical networks account for the coexistence of modularity, local clustering and scale-free topology in real systems. The traditional attributes of *network topology* are depicted in Fig. 5.1. Typically, nodes feature tree architecture, while hubs have star architecture.

The reverse engineering of biochemical networks (uncovering gene or gene regulatory function) is a central problem in SB. In recent years several methods have been developed for this purpose, using techniques from a variety of fields [3]. To date, few cellular networks have been reconstructed and analyzed in full. Examples include some prokaryotes and a few eukaryotes, such as *S. cerevisiae*. Important network properties include hubs, modularity, motifs and clusters, and redundancy. Lee et al. [4] proposed six basic network motifs in the yeast transcriptional regulatory network, representing the simplest units of the network architecture (auto-regulatory, feed-forward, single input, multi-input, etc.).

There has been considerable attention paid in recent years to network motifs, which are characteristic network patterns, or subgraphs, in biological networks that appear more frequently than expected given the degree distribution of the network [5]. Such subgraphs have been found to be associated with desirable (or undesirable) biological function (or dysfunction). It is now commonly understood that motifs constitute the basic building blocks of cellular networks [5, 6].

Pathway maps and process ontologies tools, as well as network mining tools have been listed in Nikolsky et al. [7]. More systematic effort is need to develop metabolic pathway, signaling pathway, protein interaction and gene regulation databases, for querying, visualization and analysis, in a standard exchange formats to allow their integration on a large scale [8].

5.2 Reconstructing Gene Networks

While forward genetics aims to identify mutations that produce a certain pheno-type (organisms or animal) reverse genetics seeks to determine the phenotype that results from mutating a given gene. Reverse engineering of genetic regulatory networks (GNR) from experimental data is the first step toward the modeling of genetic networks [9].

A growing number of quantitative tools for reverse engineering of GRN have been reported [10]. While a decomposition strategy to break down the entire network into a set of unit subnetworks is typically adopted, due to a very limited number of data points, this often means that for one to infer a whole GRN one must perform several separate regressions.

Several methods of analysis of time-series of gene expression exist: continuous spline function method [11], autogressive equations method [12] and hidden Markov models [13], the latter two to analyze gene expression dynamics. Others have modeled such data using mechanistic modeling via differential equations [14], the dynamic Bayesian networks approach [15] and singular value decomposition [16]. Ernst et al. [17] proposed an input–output hidden Markov model. A similar approach can be adopted for signaling pathways (e.g., for mechanistic modeling see [18]).

The methods currently used to analyze single database sets (e.g., only genomic) are quite mature and refined. Typical algorithms are those of ARACNe (([19–21] or from [20]). To infer network models that describe how a network responds to stimuli, as well as through what molecular interactions and mechanisms in this sensing and response occurs, comprehensive response profiles must be measured following perturbations. Typically, three methods to perturb the system are: RNAi, drugs, and natural variation. However, tools for integration of different platforms and approaches are still in their infancy [1].

5.3 Data Mining and Heuristic Data Preprocessing Tools

In general, data mining is the process of extracting hidden patterns from data. As more data is gathered, with the amount of data doubling every 3 years, data mining is becoming an increasingly important tool for transforming this data into information. In the area of human genetics, the goal is to understand the mapping relationship between the inter-individual variation in human DNA sequences and variability in disease susceptibility.

Heuristic (from Greek "discover") is an adjective for experience-based techniques that help in problem solving, learning and discovery. A heuristic method is typically used to rapidly come to a solution that is hoped to be close to the best possible answer, or 'optimal solution'. Heuristics are "rules of thumb" (ROT), educated guesses, intuitive judgments or simply common sense and are a general way of solving a problem. SB and BI data analysis toolboxes need general-purpose, fast and easily interpretable *preprocessing tools that perform data reduction and integration during exploratory data analysis, prior to rigorous CSB modeling.* The data mining technique that is used to perform this task is known as multifactor dimensionality reduction [22, 23]. Whilst accepting that heuristics provide less accurate processing of information compared to the solution of analytical equations, the intelligent choice of the simplifications coupled with the empirical verification of the resulting heuristic has proven itself to be a powerful systems modeling paradigm.

Several heuristic methods are analyzed below. The multivariate data analysis (MVDA) overcomes challenges associated with multidimensionality of the dataset, multicollinearity, missing data, and variation introduced by disturbing factors such as experimental error and noise [24]. Principal component analysis (PCA), partial least-squares (PLS), and multiple regressions are some of the commonly used projection methods.

MVDA is usually a multistep technique. A suitable set of statistical tools allows us to choose rationally between different mechanistic models of signal transduction or gene regulation networks. This is particularly challenging in SB where only a small number of molecular species can be assayed at any given time and all measurements are subject to measurement uncertainty. For the case of parameter estimation when likelihoods are intractable, approximate Bayesian computation (ABC) frameworks have been applied successfully [25].

Canonical Correlation Analysis (CCA) focuses on mutual dependencies and eliminates source-specific "noise". However, it does produce a separate set of components for each source. It uses a new way for dimensionality reduction, and inherits its good properties of being simple, fast, and easily interpretable as a linear projection. CCA usefulness has been demonstrated on differentially expressed genes in leukemia. The software package is available at http://www.cis.hut.fi/projects/mi/software/drCCA/ [26]. A non-linear version of PCA has also been described [27].

5.4 Analysis of Disease 'Correlation Network™' and Concerted Metabolic Activation: Disease as a Systems Network Property

Novel insights into biology and biomedicine can be obtained by evaluating the molecular connectivity through correlation networks. This is usually achieved by monitoring the dynamics of a system, or by measuring the system responses to perturbations such as drug administration or challenge tests. In addition, cross-compartment communication and control/feed-back mechanisms can be studied via *correlation network*™ *analyses* [28]. This is sometimes referred to as 'pattern recognition'. In fact, in the process, the emergent properties are delineated as well.

There is now good evidence from BI and SB analyses that human genetic diseases can be clustered on the basis of their phenotypic similarities and that such a clustering represents true biological relationships of the genes involved [29]. Such phenotypic similarity can be used to predict, and then test for, the contribution of apparently unrelated genes to the same functional module.

This concept is now being systematically tested for several diseases [30, 31]. Moreover, BI can be used to make predictions about new genes for diseases that form part of the same phenotypic cluster. A modular view of disease genes should help the rapid identification of additional disease genes for multi-factorial diseases once the first few contributing genes (or environmental factors) have been reliably identified. Using the same concept Paolini et al. [32] proposed a global mapping of the pharmacological space.

Expanding further, new avenues for DD are available, based on network analysis. Brakhage et al. [33] mentioned that the ongoing exponential growth of DNA sequence data will lead to the discovery of many natural-product biosynthetic pathways by genome mining for which no actual product has been characterized. New technologies based on genetic engineering are available to discover otherwise silent genes. Heterologous expression of a gene cluster under the control of defined promoters can be applied. Most promising is the *activation of pathway-specific regulatory genes*, which was recently demonstrated [34]. Such genes are frequent in many secondary metabolite gene clusters. This approach is rendered feasible by the fact that all of the genes encoding the large number of enzymes required for the synthesis of a typical secondary metabolite (a typical product of the fermentation industry) are clustered and that in some cases, a single regulator controls, to a certain extent, the expression of all members of a gene cluster. The activation of gene clusters by genetic engineering will lead to the discovery of many so far unknown products and therefore represents a novel avenue for DD.

Regulatory gene activation has been also identified in mammalian systems [35]. Notch, initially discovered and well characterized in *Drosophila*, plays a key role in cell–cell communication, which involves gene regulation mechanisms that control multiple cell differentiation processes during embryonic and adult life, including timely cell lineage specification of both the endocrine and exocrine pancreas.

A novel computational approach (multicomponent fitness algorithm, which includes several statistical criteria) for revealing key transcription factors that may explain concerted expression changes in specific components of the signal transduction network by knowledge-based analysis of gene expression data with the help of gene regulatory network databases is now available [36]. ExPlain (www.biobase.de) was developed for causal interpretation of gene expression data and identification of key signaling molecules.

It is important to note that correlation network^{TM} analysis works mostly for microorganisms (including yeasts), worms and flies, where good compendium data exist. It is important that these data span all kinds of OMICs and fluxome observations. The status of similar efforts for mammalian cells is rather poor. It is

(a)

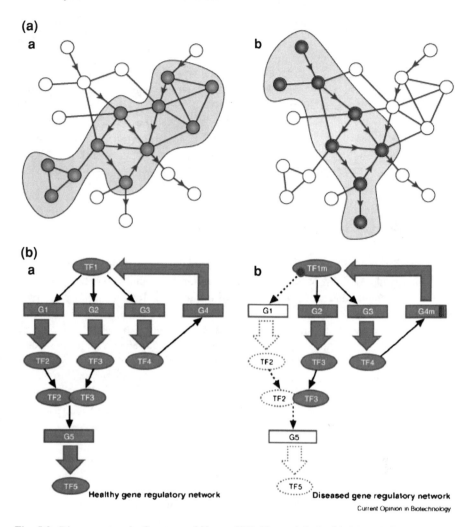

Fig. 5.2 Disease network. Courtesy of Nature [76]. Network in health (**a**) and disease (**b**). An oversimplification of network modules suggested by [76]. In disease network, some genes can be upregulated, while others are downregulated. *Blue circles*—physiological genes (products); *Red circles*—disease genes (products); *line*—bidirectional interaction; *arrow*—directed interaction. Figure 8b Disease network Courtesy of Elsevier [38]

even more problematic for cancer cell lines and other more complex, physiologically based diseases. However, extensive evolutionary conservation of orthologous genes will allow extending such computations to more important model species. So far there have been only limited attempts to specify functions of key mammalian cells and cancer genes by computational correlation network^{TM} analysis [31]. The *complex disease network* (Fig. 5.2a, b) of suitable type, preferably via OMICs, can provide a first step towards a "network-based explanation of the emergence of complex polygenic disorders" [37, 38].

On a similar note, in a first success story reported, Faratian et al. [39] have shown how the SB approach can generate hypotheses that can be tested experimentally in preclinical models and which can then be applied to clinical evaluation. Predictions from this model are consistent with known findings, and add weight to the use of PTEN as a biomarker for stratifying patients for a HER2 inhibitor or combinatorial therapy, particularly a RTK inhibitor and PI3K inhibitor in cancers with low PTEN/PI3K ratio. SB approaches, particularly deterministic kinetic models based on experimental data, offer a new approach for integrating molecular pathology and computational modeling to more rationally interrogate cancer pathways and predict responses to therapy.

5.5 Challenges for Stem Cells: Control

Stem cells are characterized by their ability to renew themselves through mitotic cell division and to differentiate into a diverse range of specialized cell types (valid for embryonic and adult stem cells). In adult organisms, stem cells and progenitor cells serve as a repair system for the body, replenishing specialized cells, but also maintaining the normal turnover of regenerative organs. Progenitor cells are already far more specific: they are developed to differentiate their "target" cell. There are some systems quantitative approaches, reviewed below, which may help research into the cell biology of stem cells.

5.5.1 SB and BI Inputs

Most adult tissues consist of stem, progenitor, and mature cells, and this hierarchical architecture may play an important role in the multistep process of carcinogenesis [40]. The authors developed and discuss the important predictions of a simple mathematical model of carcinogenesis (cancer initiation) and early progression within hierarchically structured tissue. This work presents a model that incorporates both the sequential acquisition of phenotype altering mutations and its interaction with a tissue-specific hierarchy. A novel aspect of the model is that symmetric self-renewal, asymmetric division, and differentiation are all incorporated, which enables the quantitative study of the effect of mutations that deregulate the normal, homeostatic stem cell division pattern.

Small molecule libraries have been used successfully to probe several biological systems that have potential for cell manipulation control [41]. More recently, several small molecules that control self-renewal and differentiation in stem cells have been identified. These small molecules provide useful chemical tools (i.e., probes) for both basic research and practical applications [42, 43]. The screening of stem cell control functions is the key to regenerative medicine.

Finally, additional challenges concern the construction of mathematical models to guide the design of an improved or modified stem cell niche for the desired alteration of cell fate (differentiation). Individualized treatment algorithms for regenerative medicine will feature quantification of the inherent reparative potential to identify patients could potentially benefit from stem cell therapy. This would be useful for prediction, diagnosis, prognosis, and targeting of safe and effective stem cell therapies at the earliest stage, the most natural therapeutic approach [44].

5.6 Emergent Properties

The central dogma of SB is that it is the dynamic interactions of molecules and cells that give rise to biological function (emergent property) via computational modeling to reconstruct complex systems from a wealth of reductionist, molecular data (e.g., gene/protein expression, signal transduction activity, metabolic activity, cell–cell interactions, etc.).

A number of deterministic, probabilistic, and statistical learning models are used to understand sophisticated cellular behaviors such as protein expression during cellular differentiation and the activity of signaling networks (Bhalla and Iyengar 1999). However, many of these models are bimodal. In contrast, tensor models can analyze multimodal data, which capture much more information about complex behaviors such as cell differentiation.

Many phenomena at subcellular/cellular levels, typically described in biochemistry and cell biology textbooks could be considered as phenomena resulting from interaction between different compartments and hierarchies (e.g., example, cell growth, division/proliferation, apoptosis, cell activation, homeostasis, cell death, differentiation, bacterial sporulation, system robustness, redundancy, system bistability, multiplicity of steady-states, hysteresis, oscillations, etc.). The scope of computational biological research should be to *redefine such properties in terms of mechanisms and quantify them via the SB approach.* That is, to elucidate these properties in terms of interacting species and topologies of lower levels. The systems view then dictates that one target network state resulting from dynamic molecular and physiological networks and their emergent behavior, rather than individual genes or proteins as a new strategy for DD, suggesting that we explore multitarget drugs or non-additive combination therapies (see Chap. 3).

Complexity is a property of systems with interacting parts. And when the interactions are nonlinear, it is not possible to reduce the system's behavior to a simple sum of those parts. Closely related to complexity is the concept of emergence [45]. Emergence (strong) is generally taken to mean simply that the *whole is more than the sum of its parts, or that system-level characteristics are not easily derivable from the local properties of their constituents* [46]. This implies that though higher level phenomena are not reducible to physical laws, they may still be consistent with them [45]. Therefore, the modeling of some biological processes can not solely follow a bottom-up approach, but must eventually include high-level organizing principles and even downward causality. As such, complex systems are hard to analyze using traditional mathematical and analytical methods. However, *emergence can be studied and revealed computationally* [47, 48]. Rasmussen et al. [49] have presented a simple two-dimensional molecular dynamics lattice model for interacting chemical species capable of generating higher-order emergent properties. Marsh et al. [50] have adopted a quantitative measure of EP based on information theory. The concept of emergence may serve as one of the most unifying themes across scientific disciplines, notably in biology.

In a relatively simple way, mechanistic, stoichiometric formulation allows the computation of extreme pathways that in turn enable the study of the emergent properties of the signaling network. The emergent network properties that can be analyzed in this way include [51]:

- Feasible input/output relationships,
- Crosstalk,
- Pathway redundancy,
- Reaction participations in the systemic pathways, and
- Correlated reaction sets.

These are integrated network properties that are non-obvious and impossible to derive intuitively from a simple visual inspection of reaction maps. At a higher level of organization, interaction between modules, in mathematical terms, can produce novel behavior.

Complexity and emergent properties in biology derive from several features: first, a multitude of complex inputs that stimulate multiple pathways; second, multiple outputs that present an integrated network response to inputs; third, interactions between multiple cell types; fourth, multiple environmental contexts for each cell type or combination of cells at the whole body level [52]. The immune system has been modeled as a complex system, exhibiting emergent properties [53], as well as cell size at S phase initiation (cell cycle) as an emergent property [54]. *The phenomenon of emergence is the single most important benefit of CSB.*

5.7 Computational Systems Biology

Computational systems biology (CSB), a subset of SB, aims to develop and use efficient algorithms, data structures and communication tools to allow for the integration of large quantities of biological data with the goal of modeling and simulation of the system as a whole so as to allow for the generation of testable hypotheses and pathway interrogation *in silico*. To restate this: we define "Computational Systems Biology (CSB) as quantitative, post-genomic, post-proteomic, dynamic, multi-scale physiology" [55]. In this context, the emergent property of a complex system results from the interplay of the cause-and-effect among simpler parts. In another way, CSB could be also defined as quantitative biology of function (physiology).

Traditional study of biological systems requires reductionist methods in which quantities of data are typically obtained in the form of concentrations over time in response to a certain stimuli. Computers are critical to the analysis and modeling of these data. The goal of CSB is to create accurate real-time models of a system's response to environmental and internal stimuli, such as a model of a cancer cell, in order to find targets for therapeutic intervention. Two important markup model representation languages for systems biology are the Systems Biology Markup Language (SBML) and CellML.

Elucidating complex biological networks is of central interest for understanding cellular function and the mechanisms of disease. Genetics and chemical biology have emerged as powerful techniques for dissecting cellular circuits through the controlled perturbations of protein function (stimulus–response approach), and traditional studies have been successful in elucidating the roles of core pathway components. *Interrogating biological networks, vulnerability analysis, hypothesis inference, knockdown in silico experiment, network rewiring*, etc. are typical terms used to describe such activity leading to the determination of global network properties.

Often, an in silico reconstituted network is disturbed by removal of reactants or inactivation of a poorly connected node that disrupts the specific function of the subsystem to generate hypotheses of metabolite, regulatory or signaling network function, some of which are then subsequently verified experimentally. This is in opposition to empirical (intuitive) reasoning which may be unable to grasp the complexity of the task owing to non-obvious interactions and effects.

CSB can either be approached from a top-down or bottom-up perspective. The latter is a more global attempt to identify and understand all component molecules and how they interact, whereas the former is a more focused, targeted analysis examining mechanisms. The reality is that *most often a hybrid approach is adopted* [56].

Although the emphasis formally lies on inductive discovery science, such discoveries rarely lead to molecular knowledge. These studies must either transform into or associate with the more mechanism-based, bottom-up SB. The reductionist approach may generally be most effective for acute and simple diseases, whereas a *systems approach may be most applicable to chronic and complex diseases* (arthritis, cancer, schizophrenia, metabolic syndrome and chronic pulmonary disease).

The in silico SB platforms allow cost-efficient experimentation and hypothesis exploration, computationally uncovering the behavior of molecular species that would be difficult, impossible, or too expensive to carry out in wet-lab settings. The very first hierarchical cancer apoptosis cell model has been put forward by Gene Network Sciences [57], featuring an interconnected signal transduction, gene expression network, together with cell proliferation and apoptosis. Interpolating regulatory networks from genomics data is an important solution with applications spanning all of biology and biomedicine [58].

A novel method called Biological Objective Solution Search (BOSS) for the inference of an objective function of a biological system from its underlying network stoichiometry as well as experimentally-measured state variables has been presented [59]. This new approach allows for the discovery of objectives with previously unknown stoichiometry, thus extending the biological relevance from earlier methods. This procedure could be applied to mammalian systems as well as to disease states).

Dasika et al. [60] introduced optimization-based frameworks for elucidating the input–output structure of signaling networks and for pinpointing targeted disruptions leading to the silencing of undesirable outputs in therapeutic interventions. The frameworks are demonstrated on a large-scale reconstruction of a *signaling network composed of nine signaling pathways* implicated in prostate cancer. The proposed computational frameworks can help elucidate the input/output relationships of signaling networks and help to guide the systematic design of interference strategies.

Here is a short list of steps to be performed when employing gene network reconstruction and interrogation (correlation network^{TM} inference (CNI), Chap. 8):

- Collect disease-based HTS-genomic data from the in vitro or in vivo environment (normal cells, disease cells and drug-perturbed disease cells), including transcriptomic, proteomic and metabolomic data,
- Integrate databases (thresholding, normalization, dataset alignment),
- Identify correlation substructures
- Define functional classes of data and carry out annotated functional correlation

  network^{TM} analysis (co-expression cluster, co-regulated, nodes/modules),

- Link gene ontologies with process ontologies, metabolic and signaling maps,
- Based on an association coefficient assign confidence scores (or weights) to

  each pair, both for positive and negative correlations, for functional processes or canonical maps and remove weak associations,

- Define a correlation network^{TM} (relevance network),
- Validate the significance of the correlation network^{TM} by text mining and additional biological data,
- Generate a hypothesis and potential drug targets (or biomarkers) while defining novel sites for intervention (see Figs. 8.11 and 8.13 in Chap. 8).

In this case, the *outputs of CNI are mostly qualitative* (except confidence scores), in terms of full network resolution maps. The main output is in terms of defining the network topology, particularly in targeting motifs and modules. Practical examples are in Morel et al. [61] and Yan et al. [62] and the methodology in Nikolsky et al. [63] and Ekins et al. [64, 65].

Likewise, metabolic network reconstruction methods, reaction network inference (RNI) (Chap. 8), follows similar steps to the above, but the tools are more elaborate and quantitative, with *emphasis on metabolic flux analysis*. Some details on software tools are in Plaimas et al. [66], Jamshidi and Palsson [67] and Liu and Neelamegham [68]. Besides the above applications, several tools employed and representative examples demonstrating the power of CSB are listed below:

- Fault diagnosis Boolean modeling was proposed for analysis of vulnerability of interconnected signaling pathways to assess functionality; this is often then experimentally confirmed [69].
- A mathematical model for combination therapy (for EGFR signaling network), pursuing hypothesis verification [70].
- Transient and permanent signal control (and of signal dosing) for TGF-β signaling model was studied in silico by Vilar et al. [71].
- Single-pair interactions, crosstalk and signaling pathway clustering in RAW macrophages have been studied experimentally and computationally [72].
- A computational approach to detect crosstalk among pathways based on protein interactions between pathway components (network-based method) was developed [73].
- The minimum knockout problem (solved by means Petri Net model) seeks a minimum size set of molecules whose removal (or knocking out) from a biological network makes the production of a set of molecules impossible, a problem of importance for the identification of molecular targets for therapies, especially cancer [74].

5.8 Summarizing

Systematic effort is needed to develop metabolic pathways, signaling pathways, protein interactions and gene regulation databases, for querying, visualization and analysis. The existing network models are inherently incomplete. The predictive value of any simulation depends on the quantitative empirical data that is needed to drive, validate and refine the model.

- Employ as many facets of all HTS-OMICs technologies, including the fluxosome, so as to improve the predictive power of correlation networks. Note: Part of the data will be used for model identification, and the other part for model validation.
- Change the parts, iteratively with random replacements.
- Expand on data, using more sophisticated information platforms as needed recognizing that inference and mining maybe hampered by heterogeneity of data formats.
- Unify various databases sources, formats and types so as to standardize the approach.

- Integrate over multiple knowledge pockets instead of following expensive experimental rediscovery noting that much relevant knowledge remains buried in literature [75].
- Evaluate the molecular connectivity within a system through a correlation network analysis, which represents a unique method for identifying therapeutic targets that can alter disease expression. Note: Motifs and modules should be more frequently incorporated into the network inference methods, with further emphasis on function definition of cellular parts, in relation to one another, particularly in terms of higher hierarchies [76].

References

1. Tonon G (2008) From oncogene to network addiction: the new frontier of cancer genomics and therapeutics. Future Oncol 4(4):569–577
2. Pe'er D, Hacohen N (2011) Principles and strategies for developing network models in cancer. Cell 144(6):864–873
3. Camacho D, Vera Licona P, Mendes P, Laubenbacher R (2007) Comparison of reverse-engineering methods using an in silico network. Ann N Y Acad Sci 1115:73–89
4. Lee NH (2005) Genomic approaches for reconstructing gene networks. Pharmacogenomics 6(3):245–258
5. Milo R, Shen-Orr S, Itzkovitz S, Kashtan N, Chklovskii D, Alon U (2002) Network motifs: simple building blocks of complex networks. Science 298(5594):824–827
6. Yeger-Lotem E, Sattath S, Kashtan N, Itzkovitz S, Milo R, Pinter RY, Alon U, Margalit H (2004) Network motifs in integrated cellular networks of transcription-regulation and protein–protein interaction. Proc Natl Acad Sci U S A 101(16):5934–5939
7. Nikolsky Y, Nikolskaya T, Bugrim A (2005) Biological networks and analysis of experimental data in drug discovery. Drug Discov Today 10(9):653–662
8. Cary MP, Bader GD, Sander C (2005) Pathway information for systems biology. FEBS Lett 579(8):1815–1820
9. Xiong H, Choe Y (2008) Structural systems identification of genetic regulatory networks. Bioinformatics 24(4):553–560
10. Kim CS (2007) Bayesian Orthogonal Least Squares (BOLS) algorithm for reverse engineering of gene regulatory networks. BMC Bioinf 13(8):251
11. Bar-Joseph Z, Gerber GK, Lee TI, Rinaldi NJ, Yoo JY, Robert F, Gordon DB, Barton HA (2005) Computational pharmacokinetics during developmental windows of susceptibility. J Toxicol Environ Health A 68(11–12):889–900
12. Ramoni MF, Sebastiani P, Kohane IS (2002) Cluster analysis of gene expression dynamics. Proc Natl Acad Sci U S A 99(14):9121–9126
13. Schliep A, Schönhuth A, Steinhoff C (2003) Using hidden Markov models to analyze gene expression time course data. Bioinformatics 19(Suppl 1):i255–i263
14. Chen T, He HL, Church GM (1999) Modeling gene expression with differential equations. Pac Symp Biocomput 29–40
15. Kim SY, Imoto S, Miyano S (2003) Inferring gene networks from time series microarray data using dynamic Bayesian networks. Brief Bioinform 4(3):228–235
16. Holter NS, Maritan A, Cieplak M, Fedoroff NV, Banavar JR (2001) Dynamic modeling of gene expression data. Proc Natl Acad Sci U S A 98(4):1693–1698

17. Ernst J, Vainas O, Harbison CT, Simon I, Bar-Joseph Z (2007) Reconstructing dynamic regulatory maps. Mol Syst Biol 3:74
18. Kholodenko BN (2006) Cell-signalling dynamics in time and space. Nat Rev Mol Cell Biol 7(3):165–176
19. Basso K, Margolin AA, Stolovitzky G, Klein U, Dalla-Favera R, Califano A (2005) Reverse engineering of regulatory networks in human B cells. Nat Genet 37(4):382–390
20. Altay G, Emmert-Streib F (2010) Inferring the conservative causal core of gene regulatory networks. BMC Syst Biol 28(4):132
21. Wang K, Saito M, Bisikirska BC, Alvarez MJ, Lim WK, Rajbhandari P, Shen Q, Nemenman I, Basso K, Margolin AA, Klein U, Dalla-Favera R, Califano A (2009) Genome-wide identification of post-translational modulators of transcription factor activity in human B cells. Nat Biotechnol. 27(9):829–39
22. Lou XY, Chen GB, Yan L et al (2008) A combinatorial approach to detecting gene–gene and gene-environment interactions in family studies. Amer J Human Genet 83(4):457–467
23. Yang Y, Adelstein SJ, Kassis AI (2009) Target discovery from data mining approaches. Drug Discov Today 14(3–4):147–154
24. Michelson S and Schofield T (1996) The biostatistics cookbook: The most user-friendly guide for the bio/Medical Scientist, Kluwer
25. Csilléry K, Blum MG, Gaggiotti OE, François O (2010) Approximate Bayesian Computation (ABC) in practice. Trends Ecol Evol 25(7):410–8
26. Tripathi A, Klami A, Kaski S (2008) Simple integrative preprocessing preserves what is shared in data sources. BMC Bioinf 9:111
27. Lee GW, Kim S (2008) Genome data mining for everyone. BMB Rep 41(11):757–764
28. van der Greef J (2005) Systems biology, connectivity and the future of medicine. Syst Biol (Stevenage) 152(4):174–178
29. Oti M, Brunner HG (2007) The modular nature of genetic diseases. Clin Genet 71(1):1–11
30. Chen Y, Zhang R, Song Y, He J, Sun J, Bai J, An Z, Dong L, Zhan Q, Abliz Z (2009) RRLC-MS/MS-based metabonomics combined with in-depth analysis of metabolic correlation network: finding potential biomarkers for breast cancer. Analyst 134(10):2003–2011
31. van der Greef J, Martin S, Juhasz P, Adourian A, Plasterer T, Verheij ER, McBurney RN (2007) The art and practice of systems biology in medicine: mapping patterns of relationships. J Proteome Res 6(4):1540–1559
32. Paolini GV, Shapland RH, van Hoorn WP, Mason JS, Hopkins AL (2006) Global mapping of pharmacological space. Nat Biotechnol 24(7):805–815
33. Brakhage AA, Schuemann J, Bergmann S, Scherlach K, Schroeckh V, Hertweck C (2008) Activation of fungal silent gene clusters: a new avenue to drug discovery. Prog Drug Res 66(1):3–12
34. Laureti L, Song L, Huang S, Corre C, Leblond P, Challis GL, Aigle B (2011) Identification of a bioactive 51-membered macrolide complex by activation of a silent polyketide synthase in *Streptomyces ambofaciens*. Proc Natl Acad Sci U S A 108(15):6258–63
35. Lomberk G, Urrutia R (2008) Primers on molecular pathways–notch. Pancreatology 8(2):103–104
36. Kel A, Voss N, Valeev T, Stegmaier P, Kel-Margoulis O, Wingender E (2008) ExPlain, finding upstream drug targets in disease gene regulatory networks. SAR QSAR Environ Res 19(5–6):481–494
37. Goh KI, Cusick ME, Valle D, Childs B, Vidal M, Barabási AL (2007) The human disease network. Proc Natl Acad Sci U S A 104(21):8685–8690
38. del Sol A, Balling R, Hood L, Galas D (2010) Diseases as network perturbations. Curr Opin Biotechnol 21(4):566–571
39. Faratian D, Goltsov A, Lebedeva G, Sorokin A, Moodie S, Mullen P, Kay C, Um IH, Langdon S, Goryanin I, Harrison DJ (2009) Systems biology reveals new strategies for personalizing cancer medicine and confirms the role of PTEN in resistance to trastuzumab. Cancer Res 69(16):6713–6720

40. Ashkenazi R, Gentry SN, Jackson TL (2008) Pathways to tumorigenesis–modeling mutation acquisition in stem cells and their progeny. Neoplasia 10(11):1170–1182
41. Emre N, Coleman R, Ding S (2007) A chemical approach to stem cell biology. Curr Opin Chem Biol 11(3):252–258
42. Schugar RC, Robbins PD, Deasy BM (2008) Small molecules in stem cell self-renewal and differentiation. Gene Ther 15(2):126–135
43. Fang YQ, Wong WQ, Yap YW, Orner BP (2007) Stem cells and combinatorial science. Comb Chem High Throughput Screen 10(8):635–651
44. Nelson T, Behfar A, Terzic A (2008) Stem cells: biologics for regeneration. Clin Pharmacol Ther 84(5):620–623
45. Davies PCW (2004) Emergent biological principles and computational properties of the Universe. Complexity 10:11–15
46. Fromm J (2005) Types and forms of emergence. http://arxiv.org/abs/nlin.AO/0506028 (June 13)
47. Buchanan M (2006) Nexus. The Groundbreaking Science of Networks. Norton WW, New York 2nd edition
48. Richards K, Bithell M, Dove M, Hodge R (2004) Discrete-element modelling: methods and applications in the environmental sciences. Philos Transact A Math Phys Eng Sci 362(1822):1797–1816
49. Rasmussen S, Baas NA, Mayer B, Nilsson M (2001) Defense of the ansatz for dynamical hierarchies. Artif Life 7(4):367–373
50. Marsh AG, Zeng Y, Garcia-Frias J (2005) The expansion of information in ecological systems: Emergence as a quantifiable state. Ecol Inf 1:107–116
51. Papin JA, Palsson BO (2004) Topological analysis of mass-balanced signaling networks: a framework to obtain network properties including crosstalk. J Theor Biol 227(2):283–297
52. Butcher EC, Berg EL, Kunkel EJ (2004) Systems biology in drug discovery. Nat Biotechnol 22(10):1253–1259
53. Ahmed E, Hashish AH (2006) On modelling the immune system as a complex system. Theor Biosci 124(3–4):13–18
54. Barberis M, Klipp E, Vanoni M, Alberghina L (2007) Cell size at S phase initiation: an emergent property of the G1/S network. PLoS Comput Biol 3(4):e64
55. Wikswo JP, Prokop A, Baudenbacher F, Cliffel D, Csukas B, Velkovsky M (2008) Engineering challenges of BioNEMS: the integration of microfluidics, micro- and nanodevices, models and external control for systems biology. IEE Proc Nanobiotechnol 153(4):81–101
56. Bruggeman FJ, Westerhoff HV (2007) The nature of systems biology. Trends Microbiol 15(1):45–50
57. Christopher R, Dhiman A, Fox J, Gendelman R, Haberitcher T, Kagle D, Spizz G, Khalil IG, Hill C (2004) Data-driven computer simulation of human cancer cell. Ann N Y Acad Sci 1020:132–153
58. Bonneau R (2008) Learning biological networks: from modules to dynamics. Nat Chem Biol 4(11):658–664
59. Gianchandani EP, Oberhardt MA, Burgard AP, Maranas CD, Papin JA (2008) Predicting biological system objectives de novo from internal state measurements. BMC Bioinf 9:43
60. Dasika MS, Burgard A, Maranas CD (2008) A computational framework for the topological analysis and targeted disruption of signal transduction networks. Biophys J 91(1):382–398
61. Morel NM, Holland JM, van der Greef J, Marple EW, Clish C, Loscalzo J, Naylor S (2004) Primer on medical genomics. Part XIV: Introduction to systems biology—a new approach to understanding disease and treatment. Mayo Clin Proc 79(5):651–658
62. Yan L, Karatsoreos I, Lesauter J, Welsh DK, Kay S, Foley D, Silver R (2007) Exploring spatiotemporal organization of SCN circuits. Cold Spring Harb Symp Quant Biol 72:527–541
63. Nikolsky Y, Ekins S, Nikolskaya T, Bugrim A (2005) A novel method for generation of signature networks as biomarkers from complex high throughput data. Toxicol Lett 158(1):20–29

64. Ekins S, Mestres J, Testa B (2007) In silico pharmacology for drug discovery: Methods for virtual ligand screening and profiling. Br J Pharmacol 152(1):9–20
65. Ekins S, Nikolsky Y, Bugrim A, Kirillov E, Nikolskaya T (2007) Pathway mapping tools for analysis of high content data. Methods Mol Biol 356:319–350
66. Plaimas K, Mallm JP, Oswald M, Svara F, Sourjik V, Eils R, König R (2008) Machine learning based analyses on metabolic networks supports high-throughput knockout screens. BMC Syst Biol 24(2):67
67. Jamshidi N, Palsson BØ (2008) Top-down analysis of temporal hierarchy in biochemical reaction networks. PLoS Comput Biol 4(9):e1000177
68. Liu G, Neelamegham S (2008) In silico biochemical reaction network analysis (IBRENA): a package for simulation and analysis of reaction networks. Bioinformatics 24(8):1109–11
69. Abdi A, Tahoori MB, Emamian ES (2008) Fault diagnosis engineering of digital circuits can identify vulnerable molecules in complex cellular pathways. Sci Signal 1(42):10
70. Araujo RP, Petricoin EF, Liotta LA (2005) A mathematical model of combination therapy using the EGFR signaling network. Biosystems 80(1):57–69
71. Vilar JM, Jansen R, Sander C (2006) Signal processing in the TGF-beta superfamily ligand-receptor network. PLoS Comput Biol 2(1):e3
72. Natarajan M, Lin KM, Hsueh RC, Sternweis PC, Ranganathan R (2006) A global analysis of cross-talk in a mammalian cellular signalling network. Nat Cell Biol 8(6):571–580
73. Li L, Yu M, Jason RD, Shen C, Azzouz F, McLeod HL, Borges-Gonzales S, Nguyen A, Skaar T, Desta Z, Sweeney CJ, Flockhart DA (2008) A mixture model approach in gene-gene and gene-environmental interactions for binary phenotypes. J Biopharm Stat 18(6):1150–1177
74. Ruths DA, Nakhleh L, Iyengar MS, Reddy SA, Ram PT (2006) Hypothesis generation in signaling networks. J Comput Biol 13(9):1546–1557
75. Ananiadou S, Kell DB, Tsujii J (2006) Text mining and its potential applications in systems biology. Trends Biotechnol 24(12):571–579
76. Barabási AL, Gulbahce N, Loscalzo J (2011) Network medicine: a network-based approach to human disease. Nat Rev Genet 12(1):56–68

Chapter 6
Development: In Vivo Pharmacology—
Systems Biology in Health and Disease II

This chapter covers qualitative in vivo approaches in animals and man, which will help to develop in silico pharmacology and PK positions. Additionally, we cover RNA interference in this chapter even though it is largely an in vitro method for characterizing the dynamics of cell physiology. And though in silico pharmacology is only in a rudimentary state, it is vitally important for clinical model based drug design (MBDD) development (see Chap. 10).

Definitions

- **Pharmacology** is the study of drug action, of interactions that occur between a living organism and exogenous chemicals that alter normal biochemical function. Substances having medicinal properties are then considered pharmaceuticals. Pharmacology deals with how drugs interact within biological systems to affect function.
- **Pharmacokinetics (PK)** studies time dependency and is a branch of pharmacology. PK is often exploited with pharmacodynamics (PD). PK includes the study of the mechanisms of absorption and distribution of an administered drug, the rate at which a drug action begins and the duration of the effect, the chemical changes of the substance in the body (e.g. by enzymes) and the effects and routes of excretion of the metabolites of the drug (ADME).
- **ADME** is an acronym in PK and pharmacology for absorption, distribution, metabolism, and excretion, and describes the disposition of a pharmaceutical compound within an organism. Often, toxicology is included (ADMET).
- **Pharmacodynamics (PD)** explores what a drug does to the body, whereas PK explores what the body does to the drug.
- **Disease model** is a pathological condition resembling a human disease that either develops spontaneously or is induced by distinct manipulations (surgery, antigen, or toxin). It can be also accessed or formulated as a virtual model.

A. Prokop and S. Michelson, *Systems Biology in Biotech & Pharma*,   69
SpringerBriefs in Pharmaceutical Science & Drug Development,
DOI: 10.1007/978-94-007-2849-3_6, © The Author(s) 2012

- **Xenobiotic receptors** are orphan receptors (such as PXR and CAR) that have been established as species-specific, regulating the expression of Phase I and II enzymes and drug transporters.
- **RNA interference** is a post-transcriptional gene silencing method, mediated by short fragments of double-stranded RNA.
- **Conditional gene knockout** is a technique that allows eliminating a specific target gene from a single organ in the body.
- **Pharmacogenomics** is the branch of pharmacology which deals with the influence of genetic variation on drug response in patients by correlating gene expression or single-nucleotide polymorphisms (SNP) with a drug's efficacy or toxicity.
- **Phenotyping and genotyping** represent analysis of phenotype and genotype, respectively.
- **In silico pharmacology** (also known as computational therapeutics and computational pharmacology) represents an extension of QSAR concept.

This section will cover phenomena at the pharmacology and PK levels, in the hierarchy of a living organism. The field encompasses drug composition and properties, interactions, toxicology, dynamics, therapy, and medical applications and anti-pathogenic capabilities. We will review animal disease models, gene knockout animal models, polymorphisms, and RNA-based modifications, at the qualitative level. Pharmacogenomics is reviewed next. Several possible quantitative tools are reviewed at the end of this section.

6.1 Animal Disease Models

Disease researchers use animal models to study different pathophysiologies and treatments. Such models must be predictable, emulating human conditions, and produce results that can be extrapolated and transposed and are often better than in vitro studies or computer models, but still have limitations. Animal models are difficult to maintain and quite costly. To develop animal models, numerous methods are used including chemical exposure, genetic 'knockout' or 'knock in', or forward genetic modeling. Particularly valuable are transgenics expressing dominant oncogenes, siRNAs and those with telomerase inserts. A special class of transgenics is called a 'reporter' animal, classified into those giving specific promoter activity, reporters of drug availability and distribution and reporters of drug activity. Typically, such animals are accessible via imaging.

The mouse is a key model organism for the study of mammalian genetics, development, physiology and biochemistry [1]. Comparative analysis of the mouse genome sequence with that of the human and other genomes has revealed a wealth of information on genome evolution in the mammalian lineage and assisted in the annotation of both mouse and human genomes. Systematic mutagenesis of the mouse genome will be an important step towards the first comprehensive functional annotation of a mammalian genome and the identification and characterization of models for the study of human genetic disease.

The culmination of decades of research on humanized mice is leading to advances in our understanding of human haematopoiesis, innate and adaptive immunity, autoimmunity, infectious diseases, cancer biology and regenerative medicine [2]. The development of these new generations of humanized mice will facilitate translational research in several biomedical disciplines.

6.1.1   Gene Knockout Animal Models

Genetic modification of mammalian model organisms, particularly the mouse, has the greatest potential to shed light on human development, physiology and pathology. Here, we review some of the techniques for knocking out (inactivating), mutating and knocking in (inserting) selected genes into the mouse genome [3]. Transgenesis consists of the addition of foreign genetic information to animals and specific inhibition of endogenous gene expression. Recently, these techniques have proven to be invaluable tools for target discovery, validation, and production of therapeutic proteins.

However, despite the generation of several transgenic and knockout models, obtaining relevant models still provides several theoretical and technical challenges. Indeed, genes of interest are not always available and gene addition or inactivation (often not complete) sometimes does not allow clear conclusions. The creation of xenobiotic receptor transgenic and knockout mice has provided an opportunity to dissect the transcriptional control of drug metabolizing enzymes, in addition to offering a unique opportunity to study xenobiotic receptor-mediated enzyme regulation in both drug metabolism and diseases ("Humanized" hPXR transgenic mice).

The generation of knockout animals is the method of choice to probe the function/impact of single genes, even in polygenic diseases as in cancer and autoimmunity [4] in order to better understand the mechanisms of disease development and progression. To achieve a 'conditional gene knockout' [5] a commonly used technique employs the Cre-loxP recombinase system.

6.2   Pheno- and Genotyping

Phenotyping assays of blood enzyme activities (if feasible) are generally more successful than DNA genotyping for predicting unequivocal outcomes of drug therapy in each and every patient [6]. While genotyping can, to some extent, predict drug disposition, efficacy, toxicity, and clinical outcome, success in individualized drug therapy currently appears unlikely because of the many shortcomings (e.g., ethnic differences) and complexities observed in an environmentally unconstrained and outbred human population. The same can be said for transcriptomics and proteomics, which also rely on available tumors, biopsies and excreta. Typically, a single nucleotide polymorphism (SNP) is determined as a DNA sequence variation occurring when a single nucleotide in the genome differs between members of a species. SNPs are genome-wide genetic markers which can reveal disease-relevant pathways and possible new targets. In addition, there is a hope that HTS methods will reveal information on the distribution of these polymorphic alleles in the target population and will enable the broad characterization of the drug candidate in in vitro systems. There are a number of potential fields in the therapy of major medical conditions in which genotyping (or phenotyping) of genetically polymorphic drug-metabolizing enzymes (DMEs) might be beneficial for drug safety or therapeutic outcome [7].

6.3 RNA Interference

RNA interference (RNAi) is a system within living cells that helps determine which genes are active and how active they are. Two types of small RNA molecules, such as microRNA (miRNA) and small interfering RNA (siRNA) are central to RNA interference. The selective and robust effect of RNAi on gene expression makes it a valuable research tool, both in cell culture and in living organisms because synthetic double strand RNA (dsRNA) introduced into cells can induce suppression of specific genes of interest.

RNAi may also be used for large-scale interference screens that systematically shut down each gene in the cell, and which can help identify the components necessary for a particular cellular process or an event such as cell division to occur. This is primarily due to siRNA's ability to silence previously un-druggable targets in the cytoplasm. This approach tremendously (by number) extends the concept of druggable targets as defined above (Chap. 2). The number of potential small molecular intervention sites, although unknown, is likely to be quite large.

Although therapeutic nucleotides and nucleosides (e.g., ribozyme and antisense oligodeoxynucleotides) that interfere with nucleic acid metabolism and DNA polymerization have been successfully used as anticancer and antiviral drugs, they often produce toxic secondary effects related to dosage and continuous use. Immunostimulatory oligodeoxynucleotides represent the most successful group of therapeutic oligonucleotides in the clinic. A newer group of therapeutic oligonucleotides, the aptamers, is rapidly advancing towards early detection and treatment alternatives that have commercial interest [8].

Despite the very high in vitro efficiency of small interfering RNAs (siRNAs) they present issues with intracellular target accessibility, specificity and delivery. They are likely to be developed into reagents for treating cancer and viral diseases in the near future [9].

Delivery strategies for siRNA become the main hurdle that must be resolved prior to the full-scale clinical development of siRNA therapeutics. Several delivery strategies for synthetic siRNA, focusing on targeted approaches show potential to become a useful and efficient tool in cancer therapy [10]. Attention will also be given to how RNA interference (RNAi) may become the method of choice to perform both target validation and identification within the industry. The use of RNAi-targeted biomarkers for proof of MoA is becoming an important tool in validation efforts in the preclinical phase of DD, helping to reduce the attrition rate of candidate drugs once they have entered the clinic [11].

6.3.1 BI Inputs

At present, computational siRNA design markedly reduces the costs of reagents and labor in early stage pharmaceutical research. Improvements in our ability to predict cross-silencing and immunostimulatory activities will facilitate the progression of siRNA-based drugs to clinical trials [12, 13]. Computational optimization may encompass target sequence input, generation of reverse complement and simulation of space guide sequences. At the same time, selection might include a rejection of off-target sequences. This approach demonstrates that BI could enhance the application potential of these tools.

6.4 Pharmacogenomics

Pharmacogenomics (PGN) operates at the intersection of the fields of pharmacology and genetics.

Utilization of PGN information has the potential of improving treatment outcomes and markedly reducing the rate of attrition of drugs in clinical development. A major gap that limits our ability to utilize PGN information in DD, DDv or clinical practice is that we often do not know the genetic variants responsible for inter-individual differences in drug metabolism or drug response. Several emerging genomic methods that can fill this gap have been reviewed [14]. These methods can be used to generate new information about drug metabolism or MoA, or to identify predictors of drug response.

6.4.1 SB and BI Inputs

SB can help us understand the key issues in PGN at different levels [15]. These key issues include the associations between molecular structure and function, the correlations between genotype and phenotype, and the interactions among gene, drug, and environment. At the molecular level, the detailed features of a gene and the relationship between genetic structure and function need to be further explored.

At the cellular level, the interactions and networks among those molecules should also be further examined. Better understanding at the tissue and organism levels can help establish the correlations between genotype and phenotype. The application of BI methods in PGN and SB should allow a more profound understanding of diseases at these different levels and lead to both individualized and systems medicine. Several tools are freely available at http://sysmed.pharmtao.com. PGN marker discovery projects are increasingly incorporated into Phase II clinical trials in the hope of identifying molecular predictors of response to therapy. It may be more productive to use PGN, or other molecular data, from small Phase II clinical trials to assess the clinical utility of previously defined putative markers rather than to use the data primarily for discovery. The FDA has advocated the use of PGN as a means to spur the discovery of new biomarkers for use in DDD. Seminal to this strategy is the ground breaking work of Sheiner [16] in the development of the 'Learn and Confirm' model for drug development. Basically, Sheiner structures the development space as a response surface, dimensionalized by patient heterogeneity (as defined by accessible biological markers) and dosing schemes (both in total dosage and dosing schedule). By sampling the support space of this surface as optimally as possible, one can deduce the structure of the response and characterize the patient subpopulations in a systematic fashion. This deductive reasoning and sampling strategy forms the "Learn" phase of the process. Once clarified, an inductive hypothesis-generated study is performed to "Confirm" one's findings about the surface and its characteristics. This basic strategy, as proposed by Sheiner has formed the basis for adaptive trial design and execution in the genomic environment (see for example, the work in enrichment designs for clinical trials, Simon [17]).

6.5 In Silico Pharmacology: Future

Several quantitative tools are reviewed in this section as they relate to the future of pharmacology. In silico PK is a rapidly growing area that covers the development of techniques for using software to capture, analyze and integrate biological and medical data from many diverse sources with **SB and BI inputs**. The intent is to use of this information in the creation of computational models or simulations that can be used to make predictions, suggest hypotheses, and ultimately provide discoveries or advances in medicine and therapeutics. It also includes similarity searching, pharmacophores, homology models and other molecular modeling, machine learning, data mining, network analysis and data analysis tools that all use a computer. Ekins et al. [18] lists additional approaches as virtual ligand screening, virtual affinity profiling (ligand-based and target-based) and data visualization methods. Ekins et al. [19] lists applications, flow charts and limitations. Overall, QSAR is a well developed tool, still undergoing changes with new modern approaches.

Several BI tools (e.g., PromoLign, ReguLign, PupaSNP, etc.) can complement experimental investigations of regulatory polymorphisms, allowing investigators to interpret whether polymorphisms exist in a sequence region with predicted functional importance [20, 21].

Novel genes can also be identified in silico, while their function can be predicted and characterized by virtue of sequence homology to other known proteins. Genomic DNA sequence data can be exploited to predict target genes and their modes of regulation, as well as identify susceptible genotypes based on SNP data. In addition, gene expression profiling technologies will allow toxicologists to mine large databases of gene expression data to discover molecular biomarkers and other diagnostic and prognostic genes or expression profiles.

Although in silico pharmacology is still a way off, the approach has great potential for linking the genome and proteome to pathophysiology (http://www.physiome.org) [22, 23]. It is also important for MBDD (see Chap. 10).

6.6 Summarizing

The inability of animal models to correctly predict some human responses and toxicity is acknowledged. Large animal models of human genetic diseases will become increasingly important because of physiological similarity with humans. Conditional gene targeting to restrict gene knockout to specific cells and time scales is a powerful tool for investigating the molecular basis of diseases. The generation of genome-wide sets of conditional knockout mice will become a large project applying two strategies: gene trapping based on random integration of trapping vectors into introns leading to truncation of the transcript; gene targeting, representing the directed approach using homologous recombination. It can be expected that in the near future genome-wide sets of such mice will be available. Transgenic siRNA is becoming a simpler, cheaper and faster alternative to the gene knockout approach by homologous recombination. Computational siRNA design markedly reduces the costs of reagents and labor in early stage pharmaceutical research. For PGN, quantitative simulation demonstrated that pharmacogenomics improves clinical trial design and treatment outcome. Incorporation of a genetically guided dose adjustment strategy into a clinical trial significantly increased treatment efficacy and reduced toxicity [24]. Likewise, genetic factors regulating the disposition, mechanism of action and toxicity of many commonly used medications can be identified by several methods, including mouse genetics. A mouse haplotype-based computational genetic analysis method may accelerate the rate of discovery of these clinically-important PGN factors. Finally, in silico pharmacology is still a way off, but of importance for MBDD (see Chap. 10).

References

1. Brown SD, Hancock JM (2006) The mouse genome. Genome Dyn 2:33–45
2. Shultz LD, Ishikawa F, Greiner DL (2007) Humanized mice in translational biomedical research. Nat Rev Immunol 7(2):118–130
3. Gondo Y (2008) Trends in large-scale mouse mutagenesis: from genetics to functional genomics. Nat Rev Genet 9(10):803–810
4. Kissler S, Van Parijs L (2004) Exploring the genetic basis of disease using RNA interference. Expert Rev Mol Diagn 4(5):645–651
5. Lewandoski M (2001) Conditional control of gene expression in the mouse. Nat Rev Genet 2(10):743–755
6. Nebert DW, Jorge-Nebert L, Vesell ES (2003) Pharmacogenomics and "individualized drug therapy": high expectations and disappointing achievements. Am J Pharmacogenomics 3(6):361–370
7. Tomalik-Scharte D, Lazar A, Fuhr U, Kirchheiner J (2008) The clinical role of genetic polymorphisms in drug-metabolizing enzymes. Pharmacogenomics J 8(1):4–15
8. Zhou J, Rossi JJ (2011) Aptamer-targeted RNAi for HIV-1 therapy. Methods Mol Biol 721:355–371
9. Thiel KW, Giangrande PH (2010) Intracellular delivery of RNA-based therapeutics using aptamers. Ther Deliv 1(6):849–861
10. Blagbrough IS, Zara C (2009) Animal models for target diseases in gene therapy—using DNA and siRNA delivery strategies. Pharm Res 26(1):1–18
11. Colombo R, Moll J (2008) Target validation to biomarker development: focus on RNA interference. Mol Diagn Ther 12(2):63–70
12. Reynolds A, Leake D, Boese Q, Scaringe S, Marshall WS (2004) Rational siRNA design for interference. Nat Biotechnol 22(3):326–330
13. Patzel V (2007) In silico selection of active siRNA. Drug Discov Today 12(3–4):139–148
14. Liao G, Zhang X, Clark DJ, Peltz G (2008) A genomic "roadmap" to "better" drugs. Drug Metab Rev 40(2):225–239
15. Yan Q (2008) The integration of personalized and systems medicine: bioinformatics support for pharmacogenomics and drug discovery. Methods Mol Biol 448:1–19
16. Sheiner LB (1997) Learning versus confirming in clinical drug development. Clin Pharmacol Ther 61:275–291
17. Simon R (2008) The use of genomics in clinical trial design. Clin Cancer Res 14(19):5984–5993
18. Ekins S, Mestres J, Testa B (2007) In silico pharmacology for drug discovery: applications to targets and beyond. Br J Pharmacol 152(1):21–37
19. Ekins S, Mestres J, Testa B (2007) In silico pharmacology for drug discovery: methods for virtual ligand screening and profiling. Br J Pharmacol 152(1):9–20
20. Johnson AD, Wang D, Sadee W (2005) Polymorphisms affecting gene regulation and mRNA processing: broad implications for pharmacogenetics. Pharmacol Ther 106(1):19–38
21. Fielden MR, Matthews JB, Fertuck KC, Halgren RG, Zacharewski TR (2002) In silico approaches to mechanistic and predictive toxicology: an introduction to bioinformatics for toxicologists. Crit Rev Toxicol 32(2):67–112
22. Noble D (2003) The future: putting Humpty-Dumpty together again. Biochem Soc Trans 31(Pt1):156–158
23. Michelson S (2006) The impact of systems biology and biosimulation on drug discovery and development. Mol Biosyst 2(6–7):288–291
24. Guo Y, Shafer S, Weller P, Usuka J, Peltz G (2005) Pharmacogenomics and drug development. Pharmacogenomics 6(8):857–864

Chapter 7
Development: Pharmacokinetics—Systems Biology in Health and Disease III

In silico PKPD/ADMET and biochemical-mechanistic methods will become a standard approach in the coming few years via the employment of BI and SB tools at the multiscale whole-body level. So far, the overall impact of toxicity markers on preclinical safety testing has been modest. The greatest benefit of PBPK models is they may allow for individualized health care.

Definitions

- *The post-WW2 R&D Pharma paradigm* (initially valid for the fermentation industry) involved a DD phase from nature, organismic modification by traditional selection and over-expression, rather than the chemical discovery phase described in Fig. 1.3.
- *Microdosing* is a technique for studying the drug fate *in humans* through the administration of low doses that are unlikely to produce whole-body effects, but are high enough to allow cellular response to be studied, with almost no risk of PK side effects.
- *Adaptive clinical trials* allow for modifications to the on-going trial based on either the observed data from the trial or external information, with the goal of improving the efficiency of trial design and increasing the probability of success.
- *The non-equilibrium PK model* is a phenomenon featuring a long residence time of the drug molecule on its molecular target.
- *A biomarker* can be any kind of molecule whose detection indicates a particular disease state.
- *A toxicity biomarker* is a characteristic that can be measured and evaluated as an indicator of pathological process in response to a therapeutic intervention.
- *In silico PKPD* is experimentation via simulation to represent the kinetics and dynamics of drug behavior.
- *Multiscale CSB* involves CSB integration carried out over a large span of multiple times and spatial scales.

A. Prokop and S. Michelson, *Systems Biology in Biotech & Pharma*, SpringerBriefs in Pharmaceutical Science & Drug Development, DOI: 10.1007/978-94-007-2849-3_7, © The Author(s) 2012

First, this section will discuss microdosing and adaptive design, the latter at the quantitative level. The non-equilibrium model of pharmacological activity may shift R&D focus to different areas while toxicity biomarkers and associated *in silico* ADMET prediction is becoming an important tool to guide the DD process at the organism level. Finally, physiology-based PK (PBPK) modeling is discussed, mathematically describing all physical and physiological processes.

7.1 Microdosing in PK

Microdosing will permit smarter candidate selection by taking investigational drugs into humans earlier. This is a purely experimental technique. Microdosing depends on the availability of two ultrasensitive techniques of positron emission tomography (PET) (PD information) and accelerator mass spectrometry (AMS). Microdosing allows for safer human studies as well as reducing the use of animals in preclinical toxicology. A direct test of linearity between microdoses and therapeutic effect should be sought as well as between the microdose and therapeutic doses at a later stage.

Microdosing studies are used to select drug candidates for Phase I clinical trials on the basis of their PK properties, using subpharmacologic doses (maximum 100 mμg) [1, 2]. Data to support the utility of microdosing are beginning to emerge, as 15 of 18 reported drugs demonstrated linear PK within a factor of 2 between a microdose and a therapeutic dose. The goal of a microdosing study is to assess human exposure in order to extrapolate the PK of higher, clinically more relevant doses. This strategy allows early evaluation of systemic clearance and oral bioavailability as well as sources of intersubject variability and questions of specific metabolite formation.

7.1.1 BI Inputs

Dose-related target activation and recognition of opposite thresholds between beneficial and toxic effects requires elucidation of underlying events [3]. Improved information on drug logistics and target PK enables effective drug selection, dose determination and prediction.

7.2 Adaptive Trial Design

Adaptive designs (Bayesian) promise the flexibility to redesign clinical trials at interim stages [4]. This flexibility would provide greater efficiency in DD [5]. Based on modifications, adaptive designs can be classified into three categories:

prospective, concurrent (ad hoc), and retrospective. An adaptive design allows modifications to be made to ongoing trials and/or the statistical procedures of ongoing clinical trials. In other developments, enrichment designs (ED) [6, 7] and randomized discontinuation designs (RDD) [8] have been proposed as a means for using personalized response profiles to adaptively align a therapy and dosing schedule to the therapeutic needs of a target subpopulation (e.g., "responders"). For example, there is a vast literature in the clinical realm aimed at subpopulaton segmentation using both molecular and physiological readouts in pain management and migraine prophylaxis (see, for example [9]).

7.3 Equilibrium Versus Non-Equilibrium PK Models

The importance of determination of binding kinetic profiles for efficacy and selectivity assessment in the context of lead discovery is well appreciated by Pharma, even though increasing evidence exists that renders compounds that exhibit long binary complex residence times are better optimization candidates. The enzymological principle of so-called slow k (off) compounds is well recognized and that efficacious compounds display a distinct kinetic signature. In standard equilibrium models of drug action, the pharmacological activity of a drug is dependent on the establishment of an equilibrium between the free concentration of the drug and the concentration of the drug bound to its pharmacological target receptor, while in the non-equilibrium model the pharmacological activity of a drug is most dependent on a departure from equilibrium conditions with respect to the drug and its pharmacological target receptor.

Copeland et al. [10] suggested that the most crucial factor of sustained drug efficacy in vivo is not the apparent affinity of the drug for its target *per se*, but rather the residence time of the drug molecule on its molecular target. A simple residence time for the binary complex between receptor and ligand is referred to as the period for which the receptor is occupied by a ligand. A slow dissociation of the drug from the primary target (intracellular receptors) continues to result in nearly complete retention of this ligand for a long period of time. One disadvantage of such long residence times is that they may enhance toxicities due to collateral drug binding.

A multitude of weak, or transient, biological interactions (dissociation constant: $K_{(d)} >$ microM), either working alone or in concert, occur frequently throughout biological systems [11]. This realization has important implications for DD as it can help to question the current paradigm of drug design to find the highest possible binders (drugs) to a given target (receptor). Development of transience can be based on several approches: employment of high-off-rates, multivalent approaches or multiple targets. The method itself requires a very strong grounding in the kinetics of enzyme reactions (*BI input*).

7.4 Toxicity Biomarkers

A biomarker often indicates a change in the expression or the state of a protein that correlates with the risk or progression of a disease, or with the susceptibility of the diseases to a given treatment. Once a proposed biomarker has been validated, it can be used to diagnose disease risk, the presence of disease in an individual, or to tailor treatments for the disease in an individual (choices of drug treatment or adminis- tration regimes). Selecting, evaluating and applying biomarkers in DD and explor- atory DDv may substantially shorten the time to reach a critical decision point [12]. Biomarkers are most useful in the early phase of clinical development when mea- surement of clinical endpoints or true surrogates may be too time consuming or cumbersome to provide timely proof of principle or dose-ranging information.

Toxicity biomarkers are only a part of a larger biomarker family. Danhof et al. [13] proposed a new classification of biomarkers. Mechanism-based PK/PD models have much improved properties for extrapolation and prediction, based on the scientific basis for rational DDD. A biomarker attempts, in a strictly quanti- tative manner, a process, which is on the causal path between drug administration and effect. The new classification system distinguishes seven types of biomarkers as indicated by authors: type 0, genotype/phenotype determining drug response; type 1, concentration of drug or drug metabolite; type 2, molecular target occu- pancy; type 3, molecular target activation; type 4, physiological measures; type 5, pathophysiological measures; and type 6, clinical ratings. As such, biomarkers should be considered at the early DD process (and along the discovery pathway), as indicated in Fig. 11.2.

It is widely perceived that the early screening of chemical entities can signif- icantly reduce the high costs associated with late stage failures of drugs due to poor ADMET properties. Drug toxic effects include, a broader sense, toxicity, mutagenicity, carcinogenicity, teratogenicity, neurotoxicity and immunotoxicity. Toxicity prediction techniques and structure–activity relationships rely on the accurate estimation and representation of physicochemical and toxicological properties. Most major organ systems have been explored through single cell type models or co-cultures. Simple cell-based assays exist for the liver, kidney, brain, immune system, heart, and co-cultures of multiple cell types from the same organ. The next major step is to develop animal and human stem cell-derived systems for major organs, expressing functional properties of the in vivo cells for predicting toxicity [14].

7.4.1 SB and BI Inputs

Understanding mechanisms of drug toxicity is an essential step toward improving drug safety testing by providing the basis for mechanism-based risk assessments [15]. Despite several decades of research on mechanisms of drug-induced toxicity

the *overall impact on preclinical safety testing has been modest*. Assessing the risk
of exposing humans to new drug candidates still depends on preclinical testing in
animals, which in many cases may not predict outcomes in humans accurately.
Targets have been identified for the application of new technologies, including *in
silico* screening, biomarkers, surrogate assays, computational toxicology and SB,
along with other emerging HTS methodologies. Powerful integrative SB software
and growing open-source data repositories offer new ways to share, reduce, and
analyze data from multiple sources.

7.5   In Silico Toxicity Prediction

In silico predictive ADMET screening of compounds is one of the fastest developing and most important areas in drug discovery [16]. To provide predictions of
compound drug-like characteristics early in modern DD decision making, computational technologies have been widely accepted for developing rapid high-
throughput *in silico* ADMET analysis with many *SB and BI inputs*.

Recent advances in technology have added several new tools to the biomarker
screening toolkit and have improved the throughput of existing quantitative assays.
Genomics, proteomics, and metabolomics have provided a wealth of data in the
search for predictive, specific biomarkers. Multiplexed ELISA assays, silicon
nanowire arrays, and patterned paper present unique abilities for fast, efficient
sample analysis over a broad dynamic range. Powerful integrative SB software and
growing open-source data repositories offer new ways to share, reduce, and
analyze data from multiple sources [17].

In the past few years, computational toxicology prediction systems have much
increased their predictive power, but still have not achieved a major breakthrough
due to lack of sufficiently large datasets covering more complex toxicological
endpoints (e.g. hepatotoxicity) [18]. *In silico* techniques for the prediction of
toxicological endpoints are extremely appealing because of their rapid and
effective return of results and their low cost. Moreover, these techniques can be
used in the very early phases of DD, even before the molecule is synthesized.
Numerous commercially available and free web-based programs for toxicity
prediction are available (see 18). One should, however, use caution: most do not
explicitly account for the toxicity of the reactive metabolites as well as of the
parent compound itself.

In silico prediction methods that are widely used in the pharmaceutical industry
can be roughly classified into so-called 'expert systems' and 'data driven systems'
[19]. Data driven systems are most commonly used to make predictions for
compounds with similar structures to those contained in the database and that most
probably produce toxicological effects through the same mechanism (e.g., data
driven QSARs). Clearly this approach is limited and highly dependent on the way
the chemical "shape" is captured in the database (Michelson, unpublished).

Several public efforts are aimed at discovering patterns or classifiers in high-dimensional bioactivity space that predict tissue, organ or whole animal toxicological endpoints. Novel tools to simulate complex chemical-toxicology data sets and to evaluate the relative performance of different machine learning methods have become available: Artificial Neural Networks (ANN); K-Nearest Neighbors (KNN); Linear Discriminant Analysis (LDA); Na Bayes (NB); Recursive Partitioning and Regression Trees (RPART); Support Vector Machines (SVM). Recently, a breakthrough has been reported, as Jenwitheesuk et al. [20] proposed and tested virtual screening of drug-like compounds simultaneously against the atomic structures of multiple protein targets (molecular modeling, see Chap. 3), taking into account protein–inhibitor dynamics.

In silico ADMET is emerging and rapidly evolving as a co-decisive discipline in pharma R&D. ADME data are important in (pre)clinical strategies and for early PK evaluation. Integration of these approaches will contribute towards building an SB tool for toxicology that will provide mechanistic understanding of the effects of chemicals on biological systems and aid in rational risk assessments.

7.6 Quantitative PKPD/Tox Modeling

PK-PD (PKPD) integration is essential for target validation, optimization and development of lead compounds (lead generation and lead optimization) and scaling these to human physiology. PKPD is nowadays discussed together with ADME/Tox (ADMET) and PBPK. A view of the process is illustrated in Fig. 7.1.

There have been considerable advances in the last few years in both the quantity and the quality of *in silico* ADMET property predictions, as most ADMET properties are now computable. Drug metabolism information is a necessary component of DDD. The key issues in drug metabolism include identifying the enzyme(s) involved in drug metabolism, the end product of that metabolic activity (e.g., glucuronidation, methylation, etc.), the site(s) of metabolism, the resulting metabolite(s), and the rate of metabolism. Methods for predicting human drug metabolism from in vitro and computational methodologies and determining relationships between the structure and the metabolic activity of molecules are also critically important for understanding potential drug interactions and toxicity.

Physiology-based pharmacokinetic compartmental (PBPK) modeling is used to describe, mathematically and in as much detail as possible, all physical and physiological processes which determine the PK of a substance. PBPK attempts to sub-divide the organism into single organs and to describe the disposition of a substance in each of the compartments obtained in terms of physical and physiological processes. The most commonly regarded processes are transport with blood flow, permeation processes (e.g. passive diffusion or active transport across the gut wall or into the cellular space of an organ), partitioning between blood (and plasma) and organ tissue, and metabolism and excretion. Several commercial software tools for PBPK modeling exist on the market [21].

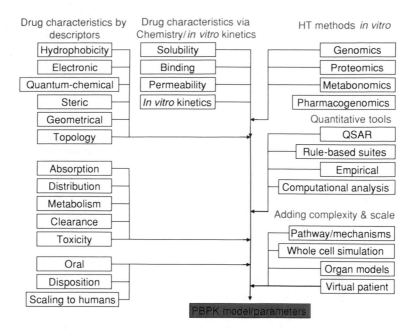

Fig. 7.1 Flow Chart of Developing ADME/Tox & PBPK. Note that feedbacks between experiments and modeling are not shown

Whole-body physiologically-based pharmacokinetic (WB-PBPK) models mathematically describe an organism as a closed circulatory system consisting of compartments that represent the organs important for compound absorption, distribution, metabolism and elimination, defined by authors [22]. Several methods are used at basic as well as clinical research at Pharma. This tool is primarily used as a means to allometrically scale PK from animals to humans based on physiology. Physiologically based PK models provide an estimate of the potential changes in internal dose that may occur throughout the life cycle [23]. These models require inputs describing changes in physiology, metabolism, and exposure with age and life stage. The growth of SB is expected to change this over the coming decade [24].

7.7 Summarizing

While microdosing is appropriate for peptide and protein therapeutics, a database will need to be built to compare PK parameters at microdosing and pharmacological doses. Furthermore, there is a need for additional biomarkers to predict toxicity in preclinical studies for making efficacy and cost-dosing decisions or terminating drug development more quickly. Rapid progress is expected at prediction of the metabolic fate of a compound. The end products of this advance will be derivation of metabolite structure and abundance which in turn will yield the concomitant insights needed in ADMET modeling. Taken together with a major effort in collecting reliable and meaningful data from general toxicity studies and, even more challenging, from clinical trials, these technologies should help to establish more reliable toxicity prediction *in silico*. To speed clinical development, a combination of *in silico* approaches with mechanism-oriented in vitro toxicity testing panels is recommended, because of failures to predict rarely occurring events (idiosyncratic toxicities). This step is facilitated via the PBPK modeling, readily now used [21]. Further refinement of the physiological and anatomical description of the organism will lead to even more complex and detailed PBPK models in the future. A comprehensive database on human and mammalian metabolic and regulatory pathways needs to be coupled with PBPK modeling effort. The PBPK modeling allows distinguishing between different groups of individuals by accounting for typical physiological differences due to age, gender or race. To individualize therapy, Bayesian dosing and covariance screening is applied. The combination of predictive physiology-based PK models with a description of the PD effect—either in a phenomenological manner or by means of detailed biochemical networks—will shift the endpoint of a simulation from a concentration versus time-curve towards PD outcome and gaining of mechanistic insights. However, it is important to note that phenomenological models are always incomplete, and that they are deliberately oversimplified condensations of the current state of knowledge (an abstract representation of the decisive relationships). In this way, though, they may represent a good starting point.

Several small and large Pharma companies are working on computer models for complex biological network models [25]. It is expected that SB will narrow the gap between data sources and data analysis. To speed up this development measurements at systems level for large set of human population is needed.

References

1. Lappin G, Garner RC (2008) The utility of microdosing over the past 5 years. Expert Opin Drug Metab Toxicol 4(12):1499–1506
2. Buchan P (2007) Smarter candidate selectionutilizing microdosing in exploratory clinical studies. Ernst Schering Res Found Workshop (59):7–27
3. Stumpf WE (2006) The dose makes the medicine. Drug Dis Today 11(11-12):550–555
4. Shen LZ, Coffey T, Deng W (2008) A Bayesian approach to utilizing prior data in new drug development. J Biopharm Stat 18(2):227–243
5. Chow SC, Chang M (2008) Adaptive design methods in clinical trials—a review. Orphanet J Rare Dis 2(3):11
6. Simon R (2008) The use of genomics in clinical trial design. Clin Cancer Res 14(19): 5984–5993
7. Hozo I, Djulbegovic B, Clark O, Lyman GH (2005) Use of re-randomized data in meta-analysis. BMC Medical Res Methodol 5:17

8. Freidlin B, Simon R (2005) Evaluation of randomized discontinuation design. JCO 23: 5094–5098

9. Halla DB, Meierb U, Dienerc H-C (2005) A group sequential adaptive treatment assignment design for proof of concept and dose selection in headache trials. Contemp Clin Trials 26:349–364

10. Copeland RA, Pompliano DL, Meek TD (2006) Drug-target residence time and its implications for lead optimization. Nat Rev Drug Discov 5(9):730–739 Erratum in: Nat Rev Drug Discov (2007) 6(3):249

11. Ohlson S (2008) Designing transient binding drugs: a new concept for drug discovery. Drug Discov Today 13(9–10):433–439

12. Kuhlmann J, Wensing G (2006) The applications of biomarkers in early clinical drug development to improve decision-making processes. Curr Clin Pharmacol 1(2):185–191

13. Danhof M, Alvan G, Dahl SG, Kuhlmann J, Paintaud G (2005) Mechanism-based pharmacokinetic–pharmacodynamic modeling—A new classification of biomarkers. Pharm Res 22(9):1432–1437

14. Guillouzo A, Guguen-Guillouzo C (2008) Evolving concepts in liver tissue modeling and implications for in vitro toxicology. Expert Opin Drug Metab Toxicol 4(10):1279–1294

15. Stevens JL (2006) Future of toxicologymechanisms of toxicity and drug safety: where do we go from here? Chem Res Toxicol 19(11):1393–1401

16. Mohan CG, Gandhi T, Garg D, Shinde R (2007) Computer-assisted methods in chemical toxicity prediction. Mini Rev Med Chem 7(5):499–507

17. Wierling C, Herwig R, Lehrach H (2007) Resources, standards and tools for systems biology. Brief Funct Genomic Proteomic 6(3):240–251

18. Muster W, Breidenbach A, Fischer H, Kirchner S, Müller L, Pähler A (2008) Computational toxicology in drug development. Drug Discov Today 13(7–8):303–310

19. Judson R, Elloumi F, Setzer RW, Li Z, Shah I (2008) A comparison of machine learning algorithms for chemical toxicity classification using a simulated multi-scale data model. BMC Bioinformatics 9:241

20. Jenwitheesuk E, Horst JA, Rivas KL, Van Voorhis WC, Samudrala R (2008) Novel paradigms for drug discovery: computational multitarget screening. Trends Pharmacol Sci 29(2):62–71

21. Schmitt W, Willmann S (2004) Physiology-based pharmacokinetic modeling: ready to be used. Drug Discov Today Technol 1(4):449–456

22. Edginton AN, Theil FP, Schmitt W, Willmann S (2008) Whole body physiologically-based pharmacokinetic models: their use in clinical drug development. Expert Opin Drug Metab Toxicol 4(9):1143–1152

23. Barton HA (2005) Computational pharmacokinetics during developmental windows of susceptibility. J Toxicol Environ Health A 68(11–12):889–900

24. Nestorov I (2007) Whole-body physiologically based pharmacokinetic models. Expert Opin Drug Metab Toxicol 3(2):235–249

25. Hopkins AL (2008) Network pharmacology: the next paradigm in drug discovery. Nat Chem Biol 4(11):682–690

Chapter 8
Development: Multiscale CSB— Simulation Tools

In order to cover bottom-up and top-down phenomena multiscale SB simulation tools should include organ-level considerations, and should be used in conjunction with multiscale modeling tools which have the ability to handle many orders of magnitude in both length and timescale. Several new R&D paradigms, based on CSB, are proposed, while some are already in the research stage. This effort will lead to virtual organ/disease models, emerging as important tools. Identifying and targeting a system's emergent properties is a major goal for coming years. This will cause a paradigm shift in R&D activity in Pharma yielding a move from population models to models of individualized medicine. The importance of multiscale CSB is underlined here as a great attention is given here in this section.

Definitions

- **Virtual organ/disease** is a computer-generated simulation (model) of organ state or disease.
- **Pharma paradigm** is a key method or algorithm used to achieve specific industry aims.
- **Population model** includes all kinds of variability within a population group in terms of disease status, including gender and metabonomics.
- **Systems genetics** relies on statistical methods, advanced computational algorithms, visualization, and high-performance computing and has a goal and potential to dissect and reassemble complex molecular and phenotypic networks in the context of natural genetic variation in a clinical setting.

A. Prokop and S. Michelson, *Systems Biology in Biotech & Pharma*, SpringerBriefs in Pharmaceutical Science & Drug Development, DOI: 10.1007/978-94-007-2849-3_8, © The Author(s) 2012

8.1 Defining CSB

More advanced modeling of complex biological systems requires the employment of multiscale models—length and timescale—which span many orders of magnitude. The level of biological organization (biological scales) includes atomic (quantum), molecular, molecular complexes, sub-cellular, cellular, multi-cell systems, tissue, organ, multi-organ systems, organism, population, and behavior [1, 2], from 1 nm to 1 m (10 orders of magnitude). Multiscale CSB seeks the development of computational models that must incorporate substantial representations of the underlying biological mechanisms from several biological scales and linkages between scales, as well as dynamic processes which span multiple time scales (from 1 μs to 10^9 s). They provide a fundamental infrastructure for predicting biological processes, diseases, and human behavior patterns. The discussion of chapter involves **SB and BI tools,** both of importance for building *in silico* PK, PD and toxicology.

Multiscale CSB requires:

- A definition of molecular species and of their pathways
- Identification of functional hierarchies and compartments and the specific molecular and modular architecture
- Characterization of the interaction between compartments (emergent properties),
- Dynamic model description
- The possibility of reducing compartment complexity
- Model validation
- Simulation tools that can handle widely different time and length scales.

Coupling models across large ranges of length- and time-scales is central to describing complex systems and biology [2]. Such coupling can be performed in hierarchical and hybrid multiscale modeling [3]. The question is how to combine detailed mechanistic model of lower elements with phenomenological elements of higher levels to link scales in models of cell/organism processes. Multiscale models couple behavior at the molecular level to that at the cellular, organ, environment level, etc. Levels thereby provide a route for calculating a quantitative description of the functioning organism, in both normal and pathological states. Spatial, temporal and spatio-temporal organization will result in emergent properties at higher levels. In a hierarchical multiscale approach, the model at the shortest length-scale is run to completion before its results are passed to the model describing the next level. One can arrange for a suitable matching of parameters at different levels. However, if there is significant feedback—that is, if changes at the larger length-scale affect behavior at the smaller length-scale—then this approach is no longer valid and one must use a hybrid or coupled multiscale approach where schemes are constructed with the physics, chemistry or biology dynamically coupled across the length and time-scales involved as suggested for cancer [4, 5].

A critical first step to approaching modeling at the systems level is to build a conceptual framework. This framework rests on current knowledge and hypotheses, and defines the levels of discretization (i.e., resolution). The modeling can be approached either from the *bottom-up*, or from the *top-down*. Bottom-up models require precise knowledge of each mechanism involved in the individual parts that will be linked together in network representations. The top-down approach dissects the available observations into the modular entities. Since a current challenge is to integrate knowledge from the fundamental physiological properties of different species, as they change with disease, with data from molecular and cellular investigation, we envision a more comprehensive *framework that accommodates both approaches* to study the complex processes of biology.

The vast majority of models that have been constructed for signaling networks are based on solution kinetics, neglecting a spatial heterogeneity. However, a compartmental model or a reaction–diffusion model is often needed when phenomena such as trafficking, transport and cellular geometry are of interest [6]. Like solution kinetics models, compartmental models are usually described by a system of ordinary differential equations (ODEs). A compartmental model treats the exchange of molecules between compartments as a flux, which can either be determined from a priori knowledge or be fitted to empirical observations, using the reaction–diffusion equation. A theoretical framework for scaling has been developed by Auffray and Nottale [7].

Since the availability of tools to quantify CSB represent the most important task of present biocomputing, we mention two basic approaches in more detail below: *Agent-based methods (ABM)* and coarse-graining methods. *Agent-based methods (ABM)* is a computational model for simulating the actions and interactions of autonomous individuals with a view to assessing their effects on the system as a whole with the objective of developing accurate methods and algorithms to cross the interface between multiple spatiotemporal scales. ABM combines elements of game theory, complex systems, emergence, and evolutionary programming. Monte Carlo Methods are applied to introduce randomness. This method provides both quantitative and qualitative approaches and is non-mathematical in nature. The models simulate the simultaneous operations of multiple agents in an attempt to re-create and predict the actions of complex phenomena, while the lower (micro) level of the system emerges at a higher (macro) level. Individual agents are typically characterized as boundedly rational, and are presumed to be acting in what they perceive as their own interests, such as reproduction, using heuristics or simple decision-making rules. To implement such models, several agent-based modeling software packages (Java based), typically in the public domain, such as SWARM (www.swarm.org), Ascape (www.brookings.edu/dynamics/models/ascape), and RePast (http://repast.sourceforge.net/).

ABM agents models exhibit "learning", adaptation, and reproduction. Most agent-based models feature:

- numerous agents specified at various scales (typically referred to as agent-granularity);
- decision-making heuristics;
- learning rules or adaptive processes;

- an interaction topology; and
- a non-agent environment.

Chavali et al. [8] applied ABM to discover an emergent behavior that arises from the immune system and discovered novel insights into immunological processes. Zhang et al. [9] introduced a multi-scale tumor modeling platform that understands brain cancer as a complex dynamic biosystem, based on ABM.

Coarse-graining (or CG computing) employs the *granularity* concept to the extent to which a system is decomposed into smaller components. CG systems consist of fewer, larger components than fine-grained systems; a coarse-grained description of a system regards large subcomponents while a fine-grained description regards smaller components of which the larger ones are composed. The main problems when developing big computational systems are that mathematical concepts appropriate at a certain level in the hierarchy of models are not generally applicable at the levels up or down in this hierarchy and new criteria for crossing the hierarchies must be developed.

CG has been used to describe *molecular dynamics*, to replace an atomistic description of a biological molecule with a lower-resolution coarse-grained model that averages or smooths away fine details, while investigating the longer time- and length-scale dynamics that are critical to many biological processes, such as the impact of lipid membranes and protein dynamics.

At the other end of the detail scale, coarse-grained and lattice models are used. Simulations of processes on long timescales (beyond about 1 microsecond) are prohibitively expensive, because they require so many time steps. Coarse graining methods have been used in molecular dynamic simulations of biological membranes. The aliphatic tails of lipids are represented by a few pseudo-atoms by gathering 2–4 methylene groups into each pseudo-atom and to examine a wide range of questions in structural biology. For example, Tozzini (2009) used coarse-grained simulation to describe events at protein dynamics that occur on different scales, from the nano- to the macroscale, spanning about 10 orders of magnitude in the space domain and 15 orders of magnitude in the time domain. McCarty et al. [10] used a first-principle multiscale modeling approach for the coarse-grained representation of polymer liquids.

Another tool is also available. Models that focus on cell state transitions and their consequences for intercellular communication—as opposed to details of intracellular biochemistry—are frequently formulated in terms of finite state automata (e.g., [11]). Authors state that by "adding a spatial aspect to automata models, cellular automata consist of grids of 'cells' that switch between states based on the states of their neighbor 'cells'. The modeler has to search for the most useful *definition of the different scales, their functional components and, in particular, of the ways information is exchanged between different scales of a model.*" Multiscale models are frequently of a hybrid type containing a combination of phenomenological elements and detailed mechanistic parts. Complex Automata, a generalization of Cellular Automata, allow for coupling of all spatial and temporal scales present in a complex system. The nature of the coupling and mutual distance on a scale separation map is a key factor to foster simulations crossing length and time scales. The application of so-called lattice-gas cellular automaton models is also advisable. This framework may also include agent-based models. Specifically, COAST has been developed as a multiscale framework coined 'Complex Automata' for modelling and simulation of complex systems [12]. The key tenet of COAST is that a multi-scale system can be decomposed into a number of single-scale Cellular Automata or agent-based models that mutually interact across the scales. Decomposition is facilitated by building a Scale Separation Map (SSM) on which each single-scale system is represented according to its spatial and temporal characteristics. Processes having well-separated scales are thus easily identified as the fundamental components of the multi-scale model [12]. Executable software is available (Simmune, run on Linux, MacOS 10, and Windows XP).

8.2 Redefining (and Discovering) Emergent Properties at Higher-Level Hierarchies

Emergent properties have readily been defined, via modeling, at subcellular and intercellular levels (see Chap. 5). Higher hierarchies are rarely included in these constructs. At this stage, emergent properties originating from higher level interactions have only been deduced rather intuitively [13]. Quantitative tools for extracting the emergent properties of such high-level hierarchies are still lacking. Several examples below exemplify the utility of such an approach.

Quaranta et al. [14] summarized recent efforts using mathematical modeling and computation to simulate cancer invasion, with a special emphasis on the tumor microenvironment. They considered cancer progression as a complex multiscale process and approached it with three single-cell-based mathematical models that examined the interactions between the tumor microenvironment and cancer cells at several scales. As a result, experiments were proposed to test the hypothesis that invasion is an emergent property of cancer cell populations adapting to selective microenvironment pressure, rather than the culmination of cancer progression producing cells with the "invasive phenotype".

Robertson et al. [15] presented a new computational framework that integrates intracellular-signaling information with multi-cell behaviors in the context of a spatially heterogeneous tissue environment which could be applied to mesoderm migration in the *Xenopus laevis* explant model. The model structure itself recapitulates many features of this process during development in humans. The simulation includes intracellular Wnt/beta-catenin signaling with an agent-based model of mesoderm migration across a fibronectin extracellular matrix substrate. The emergent cell behaviors in the simulation suggest that certain properties, particularly maintaining a delicate balance of integrin and cadherin interactions is needed to experimentally reproduce observed migratory behaviors. This model couples two different spatial scales in biology: intracellular with multicellular.

A theoretical framework for neural systems in which the dynamics are nested within a multiscale architecture has been proposed [16]. The dynamics at each scale are determined by a coupled ensemble of nonlinear oscillators, which embody the principle of scale-specific neurobiological processes. The dynamics at larger scales are 'slaved' to the emergent behavior of smaller scales through a coupling function.

8.3 Virtual Organs, Disease Models, Virtual Patient

Entelos, the leading company in this field, has various concepts and ideas that are worth noting and discussing. The ability to predict clinical efficacy *in silico* can save the pharmaceutical industry time and resources. Additionally, such an approach will result in more targeted, personalized therapies. To date, a number of *in silico* strategies using both **SB and BI inputs** have been developed to provide better information about the human response to novel therapies earlier in the drug development process. Some of the most prominent include physiological modeling of disease etiology and disease processes, analytical tools for population PD, tools for the analysis of genomic expression data, Monte Carlo simulation technologies, and predictive biosimulation. These strategies are likely to contribute significantly to reducing the failure rate of drugs entering clinical trials. Several virtual platforms are briefly discussed below.

A large-scale mathematical model, the Entelos Rheumatoid Arthritis (RA) PhysioLab platform, has been developed to describe the inflammatory and erosive processes in afflicted joints of people suffering from RA [17]. The platform represents the life cycle of inflammatory cells, endothelium, synovial fibroblasts, and chondrocytes, as well as their products and interactions. The interplay between these processes culminates in clinically relevant measures for inflammation and erosion. The RA PhysioLab platform has been used to systematically and quantitatively study the predicted therapeutic effect of modulating several molecular targets, which result in a ranking of putative drug targets and a workflow to confirm the simulations experimentally, in addition to presenting case studies of therapies directed against IL-12 and IL-15.

A skin sensitization PhysioLab has been used to construct a computer-based mathematical model of the induction of skin sensitization, in collaboration with Entelos, Inc. [18]. The biological mechanisms underlying the induction phase of skin sensitization are represented by nonlinear ordinary differential equations and refined using data from over 500 published papers. By using the model, it was determined that one key factor with a major influence on the induction of skin sensitization is TNF-α production in the epidermis. This information provides a biologically-relevant rationale for the interpretation and potential integration of diverse types of non-animal predictive data.

The Entelos Type 1 Diabetes PhysioLab platform, a dynamic large-scale mathematical model of the pathogenesis of type 1 diabetes, was used to study the effects of anti-CD40L therapy *in silico* [19, 20]. An examination of the impact of PK variability and the heterogeneity of disease progression rate on therapeutic outcome provided insights that could reconcile apparently conflicting data in the literature. Optimal treatment protocols were identified by exploring the dynamics of key pathophysiological pathways. In a similar direction, a large-scale dynamic mathematical model of the female NOD mouse was developed. In this model, virtual NOD mice are constructed by mathematically representing components of the immune system and islet beta cell physiology important for the pathogenesis of type 1 diabetes [21, 22].

A few other examples are below. Dopaminergic neurodegeneration during Parkinson disease (PD) involves identification of several pathways involved in the pathogenesis of the disease [23]. Vali et al. [24] have utilized SB to build a dynamic model for understanding and linking the various events related to PD pathophysiology.

A C57Bl/6-specific model, recalibrated for inflammatory analyte data in CD14-/- mice was proposed to elucidate altered features of inflammation in animals [25]. Mathematical modeling may provide insights into the complex dynamics of acute inflammation in a manner that can be tested in vivo using many fewer animals than has been possible previously.

We conclude with a statement from a recent report (Anonymous Pricewater-houseCoopers 2007):

> *Bioinformatics (note a correction by the authors of this Review: meant to be "Systems biology") experts aim to create a complete computer/mathematical representation of the molecular and cellular components of the human body—a "virtual" man—which can be used to simulate the physiological effects of interacting with specific targets, identify which targets have a bearing on the course of a disease and determine what sort of intervention is required (i.e. an agonist, antagonist, inverse agonist, opener, blocker etc.). However, developing such a model will require a monumental global effort far exceeding that of any similar work, e.g., the Human Genome Project.*

The virtual patient is already coming: besides Entelos, Optimata (Israel) is developing a virtual patient model, based on virtual physiology, disease and treatment modules [26, 27]. This would require a standardized modeling language as well as a well-organized web-based collaboration with international support.

8.4 Population Level Model: Towards Individualized Medicine

Personalized medicine will be the ultimate *clinical* **application of SB**, in which biological parameter variability in individuals and their statistical description in large populations (stratified patient population) can be used to interrogate the outcomes of therapeutic interventions and global patterns of disease distribution.

The differences between individuals also influence disease development and possibly optimized therapeutic interventions in individuals and populations [28]. Clearly, SB and MBDD (Chap. 10) will make these models better by focusing on the personalized aspects of that strategy that minimizes the adverse event profile while maximizing the efficacy window. PGN may enable clinicians to prospectively identify patients most likely to derive benefit from a drug, with minimal likelihood of adverse events. It is likely to transform the way clinical trials are conducted by allowing for the selection of a more homogeneous study population, thereby reducing the size and cost of a trial (see Chap. 10).

Highly perturbed gene/protein networks become building blocks for constructing systems level representations of hypothesized network models of the disease or condition being studied. Such a model can be loaded with multiple conditions to discover the unique differences or commonalities between different population groups as defined by genotypes, race, age, gender, and/or environmental conditions. There are many different modeling approaches available. Some examples are provided below.

Disease simulation models are used to conduct decision analyses of the comparative benefits and risks associated with preventive and treatment strategies [29]. Barendregt et al. [30] presented a simple generic disease model with incidence, prevalent state, and fatality and remission, and derived a set of equations that describes this disease process and that allow calculation of the complete epidemiology of a disease given a minimum of three input variables: for example asthma with age-specific prevalence, remission, and mortality.

8.5 Targeting Networks: Towards Organismic, Full-Scale Design

Several novel approaches demonstrate the power of targeting networks in analyzing full-scale behavior and disease state/progression. Again, the appearance of emergent properties, from such analyses, is the key issue. 'Disease modeling' and 'in silico' predictive methods are becoming more frequent, coupled with experimental tools [31]. Two examples are presented below.

A network model of the gene network that controls T-cell activation in humans, which is critical for the development of autoimmune diseases such as multiple sclerosis (MS), has been proposed [32]. It was established on the basis of the quantitative expression from 104 individuals of 20 genes from the immune system, extracted from the Ingenuity database (IPA) and by Bayesian inference. In the MS patient network there was an increase in the weight of gene interactions related to Th1 function and a decrease in those related to Treg and Th2 function. Based on these results, IFN-β therapy induced changes in gene interactions related to T cell proliferation and adhesion. Likewise, a new therapeutic target has been identified whose differential behavior in the MS network was not modified by therapy.

In vitro treatment with an agonist peptide modulated the T-cell activation in patients. This study illustrated how network analysis can predict therapeutic targets for immune intervention and identified a new therapeutic target for MS.

System-level 'onco-networks' have been proposed to explain cancer development and progression [33]. Analysis of the results of genome-wide experiments suggests that a cell can be induced to persist in one state or in transition between states, while the latter can be reversed when the high dimensional space of extracellular and intracellular parameters is understood. Authors postulate that as conditions change, certain cellular states (cell lines) are no longer supported, new ones emerge, and transitions (cell differentiation or death) occur. It appears that studying individual oncogenes may not be sufficient to understand cancer; rather, "onco-networks" (subsets of strongly coupled genes supporting multiple cell states) should be considered. Thus, the huge number of theoretically possible gene activity combinations of a *disease can be greatly reduced to a relatively small subset of characteristic gene activity* profiles that satisfy regulatory interaction rules. Such clusters may become drug targets.

Network Inference is often carried out by identifying groups of co-expressed genes from gene expression data using clustering or biclustering algorithms. Clustering of co-expression profiles allows us to infer shared regulatory inputs and functional pathways. However, it is an underdetermined (underconstrained, ill-posed) problem: we have many more parameters than data values to fit. Different inference methods are discussed by De Smet and Marchal [34].

A new term has been coined for the gene network correlations described above: 'Systems Genetics' (SG) [35]. By integrating a diversity of data like DNA variation, gene expression, protein–protein interaction, DNA–protein binding, and other types of molecular phenotype data, more comprehensive networks of genes both within and between tissues can be constructed to paint a more complete picture of the molecular processes underlying the physiological states associated with disease. Thus, SG seeks to understand this complexity by integrating the questions and methods of SB with those of genetics to solve the fundamental problem of interrelating genotype and phenotype in complex traits and disease. These more integrative, systems-level methods lead to networks that are predictive, provide a deeper context within which single genes operate, such as those identified from genome-wide association studies or those targeted for therapeutic intervention.

8.6 Redefining the Traditional R&D Paradigm

Having introduced all aspects which are important for cell and organism based therapies in previous sections it is now possible to propose SB Paradigm 2, a physiology-based approach (Fig. 8.1). *It includes both arms of the systems approach, top-down and bottom-up*, and stands in clear contrast to the classical reductionist view. The emphasis on human cell lines (cell systems biology) and

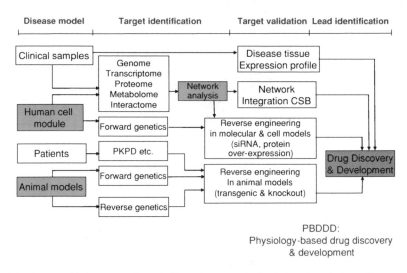

Fig. 8.1 Systems biology paradigm 2. Flow diagram of physiology-based drug discovery and development (PBDDD), presented here as a truncated standard R&D sequence, utilizing both, top-down and bottom-up routes, and both, qualitative and quantitative tools. Based on Lindsay [38]

animal models (systems pathology/pharmacology) is advocated by Butcher [36] and van der Greef and McBurney [37]. This biology-driven approach could significantly reduce the time and cost of new drug development. This paradigm is beginning to be understood and used in the industry with a good success rate. By and large, it is still *qualitative in nature*, with limited modeling and simulation, resulting in identification of important pharmaceutical leads. The approach requires substantial **BI input**. Network integration, via CSB (quantitative) represents one possible extension, if sufficient data are available. The finite window constructed by network analysis offers an unprecedented view into the molecular dynamics underlying global responses to discrete inputs. For more details see Hecker et al. [39]. One can easily visualize that Fig. 8.1 suffers from not covering a multiscale approach (cell–cell, organ and whole body) and this should be developed as soon as possible.

Another scheme is proposed with a more quantitative approach. It is a modification, a zoom-in, of Fig. 8.2, with the emphasis on 'correlation network™' inference' (CNI, Fig. 8.3). This scheme is based on employment of combinations of data from different OMICs technologies with the aim of extracting crucial systems modules/clusters (tree-based correlation networks) as possible intervention targets of drug the discovery effort. Correlation network analysis should be performed on normal cell phenotypes and compared with disease (and drug-perturbed) states. In fact, this approach integrates various interactome and functional relationship networks into a coherent and realistic context and has been applied to reveal genes potentially involved in cancer [40]. Integrating a coexpression network, seeded with four well-known breast cancer associated genes, together with genetic and physical interactions, yielded a breast cancer network model out of

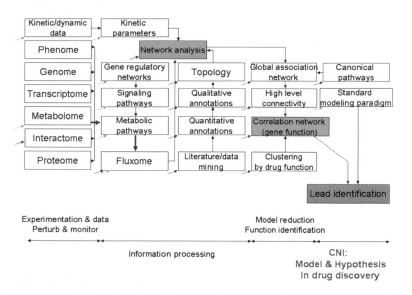

Fig. 8.2 Systems Biology Paradigm 3 Zooming in on correlation network inference for drug discovery (CNIDD), based mostly on information theory. Small blue arrows denote bioinformatics inputs, while green arrows denote chemical reaction engineering inputs (limited). The approach is largely qualitative. This scheme is very loosely based on Fig. 8.1 of Ng et al., 2006

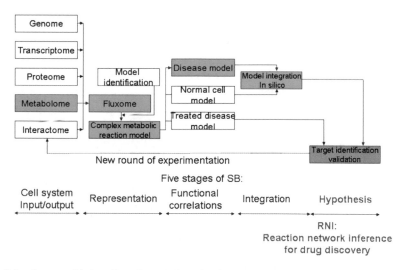

Fig. 8.3 Systems Biology Paradigm 4 Zooming-in CSB, based on detailed reaction network inference for drug discovery (RNIDD or NBDD) with emphasis on the fluxome

which candidate cancer susceptibility and modifier genes could be predicted [40]. A new term has been proposed for this strategy 'Integrative network modeling'. This strategy is equally applicable to other types of cancer and other types of disease [41, 42].

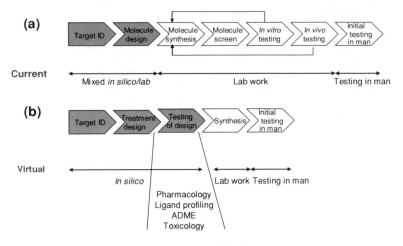

Fig. 8.4 Systems Biology Paradigm 5 Virtual testing paradigm for drug discovery and development, VTPDDD (year 2020). a *In silico* tools are currently used to design new molecules (for known targets and interactions). b Virtual Patient *in silico*. This paradigm including genetic variations and disease traits (individualized data), ADME, optimal balance between efficacy and safety and dosing. For both (a) and (b), *in silico* stages are in violet. Based on Pricewaterhous-eCoopers 2007 Report

The next level of zoom-in of Figs. 1.3 and 8.2 is the 'reaction network inference' (RNI) paradigm (Fig. 8.4) (reaction network inference for drug discovery, RNIDD, also called network-based DD—NBDD paradigm [43, 44]). This extremely ambitious effort encompasses whole cell complexity and is not likely to be achievable in the near future for significant cell system targets. It covers extremely large sets of yet undetermined components, with limited understanding of their function and interactions. The computational demand for executing such an effort is very intensive. Higher organ level will require yet further effort.

SB played an important role in understanding AstraZeneca's Iressa (gefitinib, Hendricks et al. [45]), and in identification of liver abnormalities by Pfizer [46], while Gene Network Sciences/J&J demonstrated a MoA of anticancer receptor kinase inhibitors [47]. In the neurological disease area, a coupling of biochemical networks and electrical signaling was accomplished. Swedish researchers (the Björkegren group) identified several cholesterol-responsive genes through SB-based efforts [48]. Other major companies using SB include Merck, Lilly (Singapore) and Roche, which uses *in silico* technologies in approximately 50% of its projects, mostly in Phase II and Phase III clinical trial designs [49]. It appears that most big pharma companies sponsor some level of SB research. A number of biotech companies are exploring the DD route, licensing-out, services and software focus.

A comparison between the current and virtual R&D paradigms is shown in Fig. 3.5, and the implementation of such paradigms will require both qualitative and quantitative tools (PricewaterhouseCoopers Report 2007). In vitro assays and animal disease models are often unreliable in predicting efficacy. Most promising candidates are tested in animals before forwarding them to initial testing in man. The work currently undertaken in the clinical environment can be tackled much earlier within the discovery phase.

8.7 Summarizing

Integrating models of varying complexity presents a challenging task for computing because of the many orders of magnitude they must span of time scales for elementary events. Multiscale (hierarchical) modeling is a combination of *bottom-up* and *top-down approaches* with the capability for automatic aggregation of detailed lower level models, as well as with automatic decomposition of integrated upper level models. This development will require tremendous effort to accommodate the complexity of the issues involved. Some tools are already available. In many areas of biology and physiology, multiscale models are still in their infancy [49], while CSB is much more than the conscious use of emergence. One limitation to the wide spread use of these modelling techniques in pharmaceutical R&D is the distinct lack of theory for how to integrate model selection with constraint propagation across several layers of biological organization. Developing a virtual patient model, based on virtual physiology, that must, per force, include disease and treatment modules may soon become reality, while population-level approaches that segment populations that seem to be responsive or unresponsive to a medicine are presently being developed and implemented in the clinical trial environment enrichment design [50]. These efforts typically require the identification and validation of objectively based, quantitative screens. Correlation and reaction network inference paradigms will be more effective once massive amounts of experimental data are available for mammalian systems, particularly in terms of mechanisms and reaction details. Finally, targeting the central hubs of a disease disregulated network may prove more therapeutically effective.

References

1. Prokop A (1982) Systems analysis and synthesis in biology and biotechnology. Internat J General Systems 8:1–25
2. Prokop A, Bajpai RK (1990) Bioreactor design and operation, In: progress in recombinant DNA technology and applications, Prokop A, Bajpai RK, Ho CS (eds) McGraw-Hill, New York, pp. 415–459
3. Coveney PV, Fowler PW (2005) Modelling biological complexity: a physical scientist's perspective. J R Soc Interface 2:267–280
4. Anderson AR, Weaver AM, Cummings PT, Quaranta V (2006) Tumor morphology and phenotypic evolution driven by selective pressure from the microenvironment. Cell 127(5):905–915

5. Anderson AR, Rejniak KA, Gerlee P, Quaranta V (2008) Microenvironment driven invasion: a multiscale multimodel investigation. J Math Biol 58(4–5):579–624

6. Smith AE, Slepchenko BM, Schaff JC, Loew LM, Macara IG (2002) Systems analysis of Ran transport. Science 295(5554):488–491

7. Auffray C, Nottale L (2008) Scale relativity theory and integrative systems biology: 1. Founding principles and scale laws. Prog Biophys Mol Biol 97(1):79–114

8. Chavali AK, Gianchandani EP, Tung KS, Lawrence MB, Peirce SM, Papin JA (2008) Characterizing emergent properties of immunological systems with multi-cellular rule-based computational modeling. Trends Immunol 29(12):589–599

9. Zhang L, Wang Z, Sagotsky JA, Deisboeck TS (2009) Multiscale agent-based cancer modeling. J Math Biol 58(4–5):545–559

10. McCarty J, Lyubimov IY, Guenza MG (2009) Multiscale modeling of coarse-grained macromolecular liquids. J Phys Chem B 113(35):11876–11886

11. Meier-Schellersheim M, Fraser IDC, Klauschen F (2009) Multiscale modeling for biologists. WIRE Syst Biol Med 1:4–14

12. Evans DJ, Lawford PV, Gunn J, Walker D, Hose DR, Smallwood RH, Chopard B, Krafczyk M, Bernsdorf J, Hoekstra A (2008) The application of multiscale modelling to the process of development and prevention of stenosis in a stented coronary artery. Philos Transact A Math Phys Eng Sci 366(1879):3343–3360

13. Prokop A, Davidson JM (2008) Nanovehicular intracellular delivery systems. J Pharm Sci 97(9):3518–3590

14. Quaranta V, Rejniak KA, Gerlee P, Anderson AR (2008) Invasion emerges from cancer cell adaptation to competitive microenvironments: quantitative predictions from multiscale mathematical models. Semin Cancer Biol 18(5):338–348

15. Robertson SH, Smith CK, Langhans AL, McLinden SE, Oberhardt MA, Jakab KR, Dzamba B, DeSimone DW, Papin JA, Peirce SM (2007) Multiscale computational analysis of Xenopus laevis morphogenesis reveals key insights of systems-level behavior. BMC Syst Biol 1:46

16. Breakspear M, Stam CJ (2005) Dynamics of a neural system with a multiscale architecture. Philos Trans R Soc Lond B Biol Sci 360(1457):1051–1074

17. Rullmann JA, Struemper H, Defranoux NA, Ramanujan S, Meeuwisse CM, van Elsas A (2005) Systems biology for battling rheumatoid arthritis: application of the Entelos PhysioLab platform. Syst Biol (Stevenage) 152(4):256–262

18. Maxwell G, Mackay C (2008) Application of a systems biology approach to skin allergy risk assessment. Altern Lab Anim 36(5):521–556

19. Gadkar KG, Shoda LK, Kreuwel HT, Ramanujan S, Zheng Y, Whiting CC, Young DL (2007) Dosing and timing effects of anti-CD40L therapy: predictions from a mathematical model of type 1 diabetes. Ann N Y Acad Sci 1103:63–68

20. Shoda L, Kreuwel H, Gadkar K, Zheng Y, Whiting C, Atkinson M, Bluestone J, Mathis D, Young D, Ramanujan S (2010) The Type 1 Diabetes PhysioLab Platform: a validated physiologically based mathematical model of pathogenesis in the non-obese diabetic mouse. Clin Exp Immunol 161(2):250–267

21. Whiting CC (2007) The virtual NOD mouse: applying predictive biosimulation to research in type 1 diabetes. Ann N Y Acad Sci 1103:45–62

22. Zheng Y, Kreuwel HT, Young DL, Shoda LK, Ramanujan S, Gadkar KG, Atkinson MA, Whiting CC (2007) The virtual NOD mouse: applying predictive biosimulation to research in type 1 diabetes. Ann N Y Acad Sci 1103:45–62

23. Bharath MM (2008) Insights into the effects of alpha-synuclein expression and proteasome inhibition on glutathione metabolism through a dynamic in silico model of Parkinson's disease: validation by cell culture data. Free Radic Biol Med 45(9):1290–1301

24. Vali S, Chinta SJ, Peng J, Sultana Z, Singh N, Sharma P, Sharada S, Andersen JK, Bharath MM (2008) Insights into the effects of alpha-synuclein expression and proteasome inhibition on glutathione metabolism through a dynamic in silico model of Parkinson's disease: validation by cell culture data. Free Radic Biol Med 45(9):1290–1301

25. Vodovotz Y, Chow CC, Bartels J, Lagoa C, Prince JM, Levy RM, Kumar R, Day J, Rubin J, Constantine G, Billiar TR, Fink MP, Clermont G (2006) In silico models of acute inflammation in animals. Shock 26(3):235–244

26. Kronik N, Kogan Y, Elishmereni M, Halevi-Tobias K, Vuk-Pavlović S, Agur Z (2010) Predicting outcomes of prostate cancer immunotherapy by personalized mathematical models. PLoS One 5(12):e15482

27. Agur Z (2006) Biomathematics in the development of personalized medicine in oncology. Future Oncol 2(1):39–42

28. Nicholson JK (2006) Global systems biology, personalized medicine and molecular epidemiology. Mol Syst Biol 2:52

29. Stout NK, Goldie SJ (2008) Keeping the noise down: common random numbers for disease simulation modeling. Health Care Manag Sci 11(4):399–406

30. Barendregt JJ, Van Oortmarssen GJ, Vos T, Murray CJ (2003) A generic model for the assessment of disease epidemiology: the computational basis of DisMod II. Popul Health Metr 1(1):4

31. Fattore M, Arrigo P (2005) Knowledge discovery and system biology in molecular medicine: an application on neurodegenerative diseases. In Silico Biol 5(2):199–208

32. Palacios R, Goni J, Martinez-Forero I, Iranzo J, Sepulcre J, Melero I, Villoslada P (2007) A network analysis of the human T-cell activation gene network identifies JAGGED1 as a therapeutic target for autoimmune diseases. PLoS ONE 2(11):e1222

33. Qu K, Abi Haidar A, Fan J, Ensman L, Tuncay K, Jolly M, Ortoleva P (2007) Cancer onset and progression: a genome-wide, nonlinear dynamical systems perspective on onconetworks. J Theor Biol 246(2):234–244

34. De Smet R, Marchal K (2010) Advantages and limitations of current network inference methods. Nat Rev Microbiol 8(10):717–729

35. Sieberts SK, Schadt EE (2007) Moving toward a system genetics view of disease. Mamm Genome 18:389–401

36. Butcher EC (2005) Can cell systems biology rescue drug discovery? Nat Rev Drug Discov 4(6):461–467

37. van der Greef J, McBurney RN (2005) Innovation: Rescuing drug discovery: in vivo systems pathology and systems pharmacology. Nat Rev Drug Discov 4(12):961–967

38. Lindsay MA (2003) Target discovery. Nat Rev Drug Discov 2(10):831–838

39. Hecker M, Lambeck S, Toepfer S, Van Someren E, Guthke R (2009) Gege regulatory network inference: ata integration in dynamic models–a review. Biosystems 96(1):86–103

40. Pujana MA, Han JD, Starita LM, Stevens KN, Tewari M, Ahn JS, Rennert G, Moreno V, Kirchhoff T, Gold B, Assmann V, Elshamy WM, Rual JF, Levine D, Rozek LS, Gelman RS, Gunsalus KC, Greenberg RA, Sobhian B, Bertin N, Venkatesan K, Ayivi-Guedehoussou N, Solé X, Hernández P, Lázaro C, Nathanson KL, Weber BL, Cusick ME, Hill DE, Offit K, Livingston DM, Gruber SB, Parvin JD, Vidal M (2007) Network modeling links breast cancer susceptibility and centrosome dysfunction. Nat Genet 39(11):1338–1349

41. Ergün A, Lawrence CA, Kohanski MA, Brennan TA, Collins JJ (2007) A network biology approach to prostate cancer. Mol Syst Biol 3:82

42. Lee Y, Yang X, Huang Y, Fan H, Zhang Q, Wu Y, Li J, Hasina R, Cheng C, Lingen MW, Gerstein MB, Weichselbaum RR, Xing HR, Lussier YA (2010) Network modeling identifies molecular functions targeted by miR-204 to suppress head and neck tumor metastasis. PLoS Comput Biol. 6(4):e1000730

43. Zhao S, Li S (2010) Network-based relating pharmacological and genomic spaces for drug target identification. PLoS One 5(7):e11764

44. Klipp E, Wade RC, Kummer U (2010) Biochemical network-based drug-target prediction. Curr Opin Biotechnol 21(4):511–516

45. Hendricks BS, Griffiths GJ, Benson R, Kenyon D, Lazzara M, Swinton J, Beck S, Hickinson M, Beusmans JM, Lauffenburger D, de Graaf D (2006) Decreased internalisation of erbB1 mutants in lung cancer is linked with a mechanism conferring sensitivity to gefitinib. Syst Biol (Stevenage) 153(6):457–466

46. Xu JJ, Hendriks BS, Zhao J, de Graaf D (2008) Multiple effects of acetaminophen and p38 inhibitors: towards pathway toxicology. FEBS Lett 582(8):1276–1282

47. Khalil IG, Hill C (2005) Systems biology for cancer. Curr Opin Oncol 17(1):44–48

48. Skogsberg J, Lundström J, Kovacs A, Nilsson R, Noori P, Maleki S, Köhler M, Hamsten A, Tegnér J, Björkegren J (2008) Transcriptional profiling uncovers a network of cholesterol-responsive atherosclerosis target genes. PLoS Genet 4(3):e1000036

49. Southern J, Pitt-Francis J, Whiteley J, Stokeley D, Kobashi H, Nobes R, Kadooka Y, Gavaghan D (2008) Multi-scale computational modelling in biology and physiology. Prog Biophys Mol Biol 96(1–3):60–89

50. Liu JP, Lin JR (2008) Statistical methods for targeted clinical trials under enrichment design. J Formos Med Assoc 107(Suppl 12):35–42

Chapter 9
Development: Drug Formulation and Delivery

Drug formulation and delivery is a rapidly developing field that is also a very mature area of the R&D process. Novel methods continue to contribute to improved modalities. The whole landscape will change very soon and it will have a key role to play in replacing existing drugs with expiring patents. Nanoscale delivery methods are in rapid development, and will allow for efficient cell internalization and enhanced efficacy.

Definitions

- **Drug delivery** is the method or process of administering a pharmaceutical compound to achieve a therapeutic effect.
- **Drug targeting** is the method of delivering a drug to a particular site in the organ, cell or intracellular environment, usually via attaching a ligand to a drug or delivery vehicle (nano or microsized).
- **Nanoscale delivery** employs nanosized vehicles to achieve cellular access or internalisation.
- **Formulation** is achieved by combining a pure drug with other substances to produce a final medicinal product.
- **Sustained release** is a process of drug release over time via the formulation of a drug into a vehicle which dissolves slowly and releases the active ingredient over time.

9.1 Targeting Concept and Mechanisms

Pharmaceutical companies have at least three broad strategies for protecting and enhancing their drug franchises: legal/sales and marketing defenses; chemical modification; reformulation. Reformulation strategies are more common than

A. Prokop and S. Michelson, *Systems Biology in Biotech & Pharma*,
SpringerBriefs in Pharmaceutical Science & Drug Development,
DOI: 10.1007/978-94-007-2849-3_9, © The Author(s) 2012

chemical modification strategies, and have generated significant value in life-cycle management. Here, the same basic active ingredient is used, but changes in formulation are made to improve compliance or, in some cases, efficacy. If formulation changes are modest, reformulated drugs have the advantage of following a shorter approval route than that required of new preclinical entities, needing less clinical trial data, reducing the development process to about 2–5 years.

Several strategies of drug delivery systems (DDS) are available [1]:

- PK modification has been a successful reformulation strategy for small molecules with the development of sustained-release technology.
- Changing the route of administration has also been a successful path for improving drug profiles.
- Improving bioavailability by increasing solubility or permeability, e.g. via nanoparticulate vehicles.
- Increasing the purity of the active pharmaceutical ingredient, via, e.g. the employment of specific stereoisomers.

Controlled drug delivery (CDD) has succeeded because of the emergence of three key technologies:

- Protein PEGylation.
- Active targeting to specific cells by ligands conjugated to the DDS.
- Passive targeting to solid tumors via the EPR effect (passive uptake).

A recent review article, Venkatesh and colleagues highlight the activities in the field of biomimetic systems and their application in controlled drug delivery [2]. A definition and overview of biomimetic processes is provided, with a focus on synthesis and assembly for the creation of novel biomaterials. In particular, systems are classified on the basis of three subsets, which include biological, synthetic and biohybrid approaches. Examples focus on current and proposed clinical significance for systems that mimic processes where the underlying molecular principles are well understood. Biomimetic materials have a great potential as they are exceptional candidates for various controlled drug delivery applications and have enormous potential in medicine for the treatment of disease.

9.1.1 SB and BI Inputs

The quantitative analysis of the physical, chemical and potentially biological phenomena, which are involved in the control of drug release, offers another fundamental advantage: underlying drug release mechanisms can be elucidated, which is a pre-requisite for improving the safety of a treatment and for effective trouble-shooting during production. Several empirical and mechanistic models have been suggested [3]. One of the major challenges to be addressed in the future is the combination of mechanistic theories describing drug release within delivery systems with mathematical models quantifying the subsequent drug transport within the human body in a realistic way.

9.2 Nanoscale Drug Delivery Systems

Nanoscale particles/molecules are being developed to improve the bioavailability and PK of therapeutics. Examples are liposomes, polymeric nanoparticles, nano-suspensions and polymer therapeutics. These nanomedicinal drugs feature the ability to cross biological barriers (capability of intracellular delivery and trafficking to different organelles), or passive or active targeting of tissues [4].

Although affinity targeting has many limitations, targeted DDS represent the future of therapy. A characteristic feature of nanotechnology is its ability to add new functionality to existing products, making them more competitive. Versatility is another feature, as nanotechnology has the potential to add innovative functionality to many pharmaceutical products and medical devices. Besides, this technology can deliver a combination therapy, as opposed to most commonly used drugs. Integration is the key: drug entity, gene delivery, targeting, the delivery vehicle itself and possibly a visualization agent (imaging) [5].

9.3 CSB at Formulation and Delivery

Multiscale computational modeling of DDS is poised to provide predictive capabilities for the rational design of targeted drug delivery systems, including multi-functional nanoparticles [6]. Realistic, mechanistic models can provide a framework for understanding the fundamental physicochemical interactions between drug, delivery system, and patient. Multiscale modeling, however, is in its infancy even for conventional drug delivery. The approach relies solely on **SB and BI inputs.**

The wide range of emerging nanotechnology systems for targeted delivery further increases the need for reliable in silico predictions. Several computational approaches at different scales in the design of traditional oral drug delivery systems are available. A multiscale framework for integrating continuum, stochastic, and computational chemistry models has been proposed and a several successful case studies are available [6].

Prokop and Davidson [4] identified, rather intuitively, emergent properties relevant to drug delivery in the cancer environment. These are:

- Elementary cancer metabolic and signaling quantitative models, elementary models of nanovehicular uptake, targeting, internalization and trafficking at a subcellular level.

- Model of tumor invasion and metastasis, model of capillary network growth at cellular level.
- Comprehensive PK model at tissue level.
- Comprehensive model of cancer as a systems disease, at organism/system level.

The challenge is to integrate all of the relevant knowledge and data in a systematic way and thus devise the best therapeutic and diagnostic strategies. Accurate models will require comprehensive experimental data at multiple levels of complexity. Collecting the dose–response and PD profiles in vivo from experiments encompassing perturbations for uncovering disease mechanisms will allow for hypothesis testing and verification, including high resolution noninvasive imaging of animals (and man) on targeted drug delivery within different organs by means of near infra-red (NIR) and quantitative PET imaging.

9.4 Summarizing

Many small molecule drugs are taken orally, satisfying the Ro5 are more likely to be 'blockbusters'. The emergence of target-based drugs and improvements in delivery methods may reduce off-target effects and change the present market paradigm. It is expected that experimentally validated numerical models of DD may aid in formulating drug design, selectivity and combination therapies. And while this approach could be extended to delivery vehicles with feedback properties (responding to the organism's need) future development will be required. In particular, one new feature, intracellular delivery will have to be accounted for, while nanovehicular delivery methods may change the way drugs are targeted, delivered, and internalized. Mechanisms of nanoparticle uptake have not yet been fully characterized. To advance this field, new modeling approaches that capture the inherent properties of transport dynamics (drug/carrier) and predict spatio-temporal behavior are needed [7]. It is expected that, in the near future, DDS will become a more integral component of new medicine development than the role it presently inhabits.

References

1. Fleming E, Ma P (2002) Drug life-cycle technologies. Nat Rev Drug Discov 1:751–752
2. Venkatesh S, Byrne ME, Peppas NA, Hilt JZ (2005) Applications of biomimetic systems in drug delivery. Expert Opin Drug Deliv 2(6):1085–1096
3. Siepmann J, Siepmann F (2008) Mathematical modeling of drug delivery. Int J Pharm 364(2):328–343
4. Prokop A, Davidson JM (2008) Nanovehicular intracellular delivery systems. J Pharm Sci 97(9):3518–3590

5. Schneider S, Lenz D, Holzer M, Palme K, Süss R (2010) Intracellular FRET analysis of lipid/DNA complexes using flow cytometry and fluorescence imaging techniques. J Control Release 145(3):289–296
6. Haddish-Berhane N, Rickus JL, Haghighi K (2007) The role of multiscale computational approaches for rational design of conventional and nanoparticle oral drug delivery systems. Int J Nanomedicine 2(3):315–331
7. Maly IV (2002) A stochastic model for patterning of the cytoplasm by the salutatory movement. J Theoret Biol 216:59–71

Chapter 10
Development: Preclinical Model Based Drug Development

Large scale in silico clinical development will only become a reality after some effort is exerted; some partial solutions (**SB and BI tools mentioned in previous chapters**) already exist and first attempts have been made. We acknowledge here that we are NOT in square zero and that in silico technologies have been, and are already being, used in clinical trial design and execution, used in simulation studies for adaptive trials, and used to identify patient subpopulations and markers for enrichment strategies. This goal will require a concentrated effort by all players, with considerable investment from the pharma industry and governments.

Definitions

- In silico **clinical trial** attempts to employ SB and BI tools to define a possible scenario and explore its strengths and weaknesses by simulation before the start of human trials.
- **MBDD** is model-based drug development seeking to develop a model of quantitative disease-drug-trial relationship.

10.1 Defining MBDD

Thus far, in this review, we have neglected the impact of SB and BI on clinical research and drug development. Several quantitative tools have been developed that can enable change in the current paradigm of clinical research and drug launch.

This area has been labeled as model-based drug development (MBDD) [1]. Alternatively, pharmacometrics (PM) has been used as a descriptor of these technologies [2, 3]. This effort is again viewed as a paradigm shift in DDD by encouraging a change in emphasis from drug dose-exposure to concentration-

A. Prokop and S. Michelson, *Systems Biology in Biotech & Pharma*, SpringerBriefs in Pharmaceutical Science & Drug Development, DOI: 10.1007/978-94-007-2849-3_10, © The Author(s) 2012

response. Lalonde et al. [1] suggested six stages of implementation of MBDD: trial performance matrix, quantitative decision criteria, data analysis model, PK–PD and disease model, competitor information and meta-analysis and design and trial exceution models.

MBDD is designed to influence decisions in the pharma industry by conducting quantitative analysis of PK and PD (i.e., efficacy, safety). The approach creates or confirms prior models of disease change, identifies and characterizes the placebo effect, identifies and characterizes the impact of dropouts on trial design and logistics, and measures/quantifies drug effect, to determine the value of biomarkers for a given disease or drug class to reflect changes in primary disease end points.

Continuous development of disease and safety models will form the foundation of DDD throughout the drug life cycle. Development of a new drug will start in the preclinical phase by simulating human PK/PD using in vitro tools employing human material, animal models, and any relevant prior knowledge to characterize both the compound and the metabolic pathways involved in its ADME. The clinical phase then proceeds through several learning and confirming cycles. Reference to Sheiner's early work is warranted here as a groundbreaking idea since it was the basis of modern adaptive design strategies [4]. The learning cycle starts with PK and PD-directed dose escalation followed by human proof of principle. The confirming cycle begins with patient dose finding followed by multicenter clinical trials. In the post-NDA phase, the learning cycle continues by focusing on safety, resetting benefit/risk, and developing new indications.

It is envisioned that MBDD will increase clinical productivity through more informed decisions. To facilitate corporate learning, the information generated across DDv can be systematically compiled into centralized databases to guide future drug development and regulatory decisions. The hypotheses generated via systematic modeling and simulation-based trial design should then lead to more successful trials.

The three components of MBDD are: disease, drug, and trial models. As a baseline, a healthy-state model should be added. Disease-drug-trial models seek a mathematical representation of the time course of biomarker and clinical outcomes, placebo effects, a drug's pharmacologic effects, and trial execution characteristics for both the desired and undesired responses. The importance of SB is clear as the characterization of the relationship between biomarkers (in molecular terms) and clinical outcomes are sought, as well as a linkage between preclinical and clinical biomarkers (see e.g., Allerheiligen [5], Wetherington et al. [6], Suryawanshi et al. [7] and Keizer et al. [8].

At present, various methods to develop disease-drug-trial models are available. These analyses are extremely time consuming and inherently multidisciplinary, requiring large groups of diverse experts to work together. Industry and regulatory processes are currently not organized to accommodate such collaborative research. A potential solution could be public–private partnerships to achieve common goals [3]. It is clear that we can employ previously developed and discussed tools presented in other parts of this report (Chaps. 2–8) and readily incorporate them into the suggested scheme. Some other tools and models will have to be developed afresh.

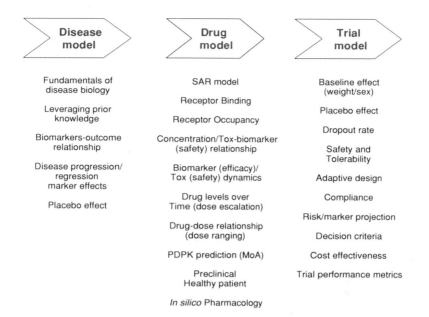

Fig. 10.1 Systems Biology Paradigm 6. Model-based drug development, MBDD: disease, drug and trial models. *Source*: Based on Gobburu and Lesko [3], Lalonde et al. [1], Powell and Gobburu [2]

As a result of the above discussion we summarize the MBDD model into another emerging paradigm (Fig. 10.1), also based on the systems approach.

Another important systems tool which could complement the MBDD is that of "Quality by design" (QbD), advocated by FDA. QbD has several attributes [9]:

- Product is designed to meet patient needs and performance requirements.
- Process is designed to consistently meet product quality attributes.
- Impact of starting raw materials and process parameters on product quality is understood.
- Critical sources of process variability are identified and controlled.
- The process is continually monitored and updated to allow for consistent quality over time.

Such programs will achieve an integration of patient needs based on robust science and clear characterization of quality requirements. The impact will be mostly felt during the development of a Pharma product and its manufacturing process, and requires a close collaboration between the industry and regulators to achieve a regulatory review based on a scientific understanding of the product and its manufacturing process. The intended result of QbD is a reduction in regulatory burden and more streamlined operations. The fruits of the program can bring a higher success rate to commercial operations with its impact on cost of goods, reduced reporting requirements, or at least a reduction in the level of reporting for supplements and less complex or stressful inspections.

   Steps in QbD typically involve (adopted from Lutze et al. [10]):

1. Define the problem (critical product attributes), where the objectives for quality by design are established.
2. Develop assays and analyses to measure the critical product attributes; incorporate process analytical technology (PAT) for monitoring of critical attributes during processing; collect all known data for the system under consideration.
3. Understand the interaction between the process and the product and identify the critical process attributes that influence product quality.
4. Develop and validate a generic model to be used for process–product analysis.
5. Generate the intensified alternatives using synthesis/design algorithms to explore the process design space.
6. Verify the alternatives trough model-based simulation.
7. Develop a statistical model that describes the process design space; optimize the feasible alternatives to identify the optimal solution.
8. Validate the optimal alternative for final selection.

10.2 Summarizing

MBDD is in an early stage of development and will require input from biotech, pharma, government and the regulatory agencies. While some components (modules) are ready to be employed (as discussed previously in this review) others will have to be developed and integrated into the overall technology suite. It is envisioned that this integrated strategy, once accomplished, will result in savings of both cost and time in clinical trials.

References

1. Lalonde RL, Kowalski KG, Hutmacher MM, Ewy W, Nichols DJ, Milligan PA, Corrigan BW, Lockwood PA, Marshall SA, Benincosa LJ, Tensfeldt TG, Parivar K, Amantea M, Glue P, Koide H, Miller R (2007) Model-based drug development. Clin Pharmacol Ther 82(1):21–32
2. Powell JR, Gobburu JV (2007) Pharmacometrics at FDA: evolution and impact on decisions. Clin Pharmacol Ther 82(1):97–102
3. Gobburu JV, Lesko LJ (2009) Quantitative disease, drug, and trial models. Annu Rev Pharmacol Toxicol 49:291–301
4. Sheiner LB (1997) Learning versus confirming in clinical drug development. Clin Pharmacol Ther 61:275–291
5. Allerheiligen SR (2010) Next-generation model-based drug discovery and development: quantitative and systems pharmacology. Clin Pharmacol Ther 88(1):135–137

6. Wetherington JD, Pfister M, Banfield C, Stone JA, Krishna R, Allerheiligen S, Grasela DM (2010) Model-based drug development: strengths, weaknesses, opportunities, and threats for broad application of pharmacometrics in drug development. J Clin Pharmacol 50(9 Suppl):31S–46S
7. Suryawanshi S, Zhang L, Pfister M, Meibohm B (2010) The current role of model-based drug development. Expert Opin Drug Discov 5(4):311–321
8. Keizer RJ, Schellens JH, Beijnen JH, Huitema AD (2011) Pharmacodynamic biomarkers in model-based drug development in oncology. Curr Clin Pharmacol 6(1):30–40
9. Nasr MM (2006) FDA, DIA meeting Philadelphia, PA, USA
10. Lutze P, Román-Martinez A, Woodley JM, Gani R (2010) A systematic synthesis and design methodology to achieve process intensification in (bio) chemical processes. In: Pierucci S, Buzzi Ferraris G (eds) 20th European symposium on computer aided process engineering—ESCAPE20, Elsevier

Chapter 11
Systems Biology: Impact on Pharma and Biotech

SB is important for DD because it can be used to rapidly identify the MoA of novel drugs, enabling companies to make go/no decisions earlier in the drug development process by avoiding pathways associated with toxicological or pharmacological issues. SB can reduce the number of compounds synthesized and manufactured owing to refined algorithms which avoid poor PK and toxic effects. In the longer term investments in SB will enable research institutions and companies to save time and money in the DD process by choosing drugs which are more likely to succeed in clinical development.

Definitions

- **Blockbuster drug:** a drug with annual sales of at least US $1 billion.
- **Niche drug** is drug focused on the (often relatively small) targetable portion (subset) of a market sector.

11.1 SB Impact

Because of stagnation in DDD, pharmaceutical R&D must adopt a basic change of paradigm in the way it does business. In short, it must speed up the process and achieve higher success rates in the clinic. SB and BI offer new tools to effectively enable this paradigm shift, by analyzing the landscape of disease, starting from observations in molecular biology and ending with manufacturing, sales and distribution. The definition of the baseline healthy state is a prerequisite. And though SB and CSB are not panaceas, i.e., they cannot possibly solve every problem that affects pharma, they can make a real contribution to drug discovery, development, and lifecycle management.

A. Prokop and S. Michelson, *Systems Biology in Biotech & Pharma*,
SpringerBriefs in Pharmaceutical Science & Drug Development,
DOI: 10.1007/978-94-007-2849-3_11, © The Author(s) 2012

"Drug development has been stagnant in terms of innovation; there exists huge potential for innovation. Failure to innovate drug development will render the "big pharma" model unsustainable"[1].

In terms of the future role of quantitative tools, which are becoming "established", in both pharmaceutical R&D and in the academic environment, it is clear that optimizing therapeutic approaches to human disease will require the application of networks (Chap. 8) to identify new drug targets (Chap. 2), and to determine the appropriate dosing of a drug based on metabolomic profiling [2]. Together these efforts will yield insights and hypotheses that will enable the investigation into the causes of resistance to therapies or enhanced toxicities. This value added knowledge will help to characterize the robustness/fragility trade-off inherent in the system. But most importantly, network biology promises to illuminate our understanding of drug action (MoA).

Advances in SB suggest that complex diseases may not be effectively treatable by interventions at single nodes of the disease network due to robust phenotypes of biological systems with compensatory signaling routes that bypass the inhibition of individual proteins. Also, recent insights from network SB predict that modulating multiple nodes simultaneously is often required to alter phenotypes. Network biology teaches us that exquisitely selective drugs may exhibit a lower efficacy than expected. Conversely, compounds that selectively act on two or more targets of interest should, in theory, become more efficacious than single-target entities.

Network analysis promises to yield multiple benefits [3]:

- SB-based network analysis can identify the determinants (nodes) or combinations of determinants that strongly influence disease expression or phenotype, providing unique insight into disease mechanism and potential therapeutic targets.
- Network analysis of disease provides the opportunity to rigorously consider relationships within the modular collection of genomic, proteomic, and metabolomic networks that interact to yield the patho-phenotype.
- Disease network analysis ultimately provides a mechanistic basis for defining phenotypic differences among individuals with the same disease through consideration of unique genetic and environmental factors that govern intermediate phenotypes contributing to disease expression.
- Disease network analysis offers a unique method of identifying therapeutic targets or combinations of targets that can alter disease expression.

The current pharmaceutical industry business model is considered to be both economically unsustainable and operationally incapable of acting quickly enough to produce the types of innovative treatments demanded by global markets. In order to make the most of these future growth opportunities, academia and industry must fundamentally change the way they operate. PricewaterhouseCoopers anticipates the following changes occurring in the industry:

- Health care will shift from treatment to prevention. The role of consciously planned nutrition and food science is increasing.
- Pharmaceutical companies will provide total health care packages.
- The current linear phase R&D process will give way to in-life testing and live licensing, in collaboration with regulators and health care providers.
- Systematic feedback and data mining for the useful and harmful effects of various drugs and drug combinations is very important.
- The traditional blockbuster sales model will disappear.
- The supply chain function will become revenue generating as it becomes integral to the health care package and enables access to new channels.

In 2004 Hood et al. [4] suggested that preventive medicine will follow disease perturbed networks to identify drug targets, first for therapy and later for prevention. This will require building a fundamental understanding of the SB that underlies normal biological and pathological processes, and the development of new technologies to achieve this goal. Predictive and preventative medicine will lead naturally to personalized medicine, which, in turn, will revolutionize health care. Drug companies will find and use more effective means of DD guided by molecular diagnostics, although the paradigm will shift to partitioning patients with a particular disease into a series of therapeutic windows, each with smaller patient populations but higher therapeutic effectiveness.

The paradigm shift is being actualized by a number of key factors:

- The phenomenal pace of technological advances, e.g. BI, combinatorial syntheses, HTS, and laboratories on a chip,
- The need for significant breakthrough discoveries
- Pressure to reduce costs
- The requirement to reduce cycle times
- Biotechnology acquisitions and mergers (survival in global markets)

This paradigm shift in design features the Innovative "experiments" that can be made *in silico* rather than in vivo or in vitro, so that only essential experiments need be undertaken.

Kola [1] states that the drug discovery process has enjoyed some significant advances, which, among others, includes diversification of the chemical compound collections, HTS for targets, important tools for target identification and validation, the use of human genetics and genetic animal models, the sequencing of the entire human genome (as well as that of several other species), and the advent/integration of new technologies such as transcriptional profiling and siRNA interference. On the other hand, the process of DDv has been relatively stagnant and pharmaceutical DDv is a highly inefficient process (one in which approximately nine molecules in ten fail during development, and many of these failures occur in the later stages). The major causes of failure in DDv have been lack of efficacy and unintended toxicity. These failures are typically due to a lack of understanding of efficacy and its proof of concept in humans and the lack of objective and robust biomarkers capable of reporting such efficacy. The evidence for the former comes partly from the observation that *compounds targeting novel mechanisms fail more frequently than those targeting network sites*, and innovation in drug development must be focused on ameliorating these two significant risks. Recent scientific and technological advances may provide the basis for that innovation. Advances in scientific technologies such as imaging, transcriptional profiling, and proteomics, provide fertile ground for exploitation of biomarker development. Rapidly evolving technologies such as functional MRI may also provide objective, robust proof-of-concept end points.

Another operational innovation is utilizing adaptive trial design more frequently. Sample size is one element that can be modified dynamically, adjusting the potential power of the trial (adding or subtracting treatment strategies, changing end points or patient populations, and altering methods of statistical analysis).

Van der Greef and McBurney et al. [5] stated: "It would be inappropriate to give the impression that the full incorporation of systems thinking into the pharmaceutical value chain will provide immediate cost savings. The implementation of such a new concept over the entire process can only take place gradually, given the existing infrastructures that might need to be changed, the current development pipelines and the regulatory constraints... Realizing the full benefits of a systems approach to drug discovery and development might take 10 years, given the infrastructures to be changed and the need to complete ongoing programmes. So, the likely way forward is a stepwise implementation based on business-driven opportunities from the clinic to discovery coupled with improvements in aspects of the process that can have widespread benefits across different therapeutic areas. Systems-based approaches provide new flexible steps into the future for improving the efficiency of drug discovery."

However, PricewaterhouseCoopers suggest a more radical approach: "We believe that incremental improvements are no longer enough; the industry will need to make a seismic shift to facilitate further progress in the treatment of disease" (PricewaterhouseCoopers 2007).

We conclude with an overview of R&D technologies as we view them emerging going forward (Fig. 11.1). They are listed in progression from those which are currently in use towards the more speculative ones (ones that still to be developed and operationalized in full). A reaction-mechanism-based tool is readily available for prokaryotes and will become standard for eukaryotes once more complete data are available. A Correlation network^{TM}-based tool will become a reality in few years, as well as quantitative clinical development tools. Virtual man (VTDD) will become commonplace in the 2020's.

Middle-ground models are forthcoming:

> ..."Between the extremes of network models and atomistic simulation a spectrum of models has been developed that might ultimately be able to bridge the daunting gaps of spatial and temporal scale. ...they also need to be coarse-grained enough to handle organism-wide processes with computational efficiency"[6]

Thus our adopted definition of Systems Biology is a two-directional, integrative, top-down (mechanism-based) and bottom-up (hypothesis-driven), often

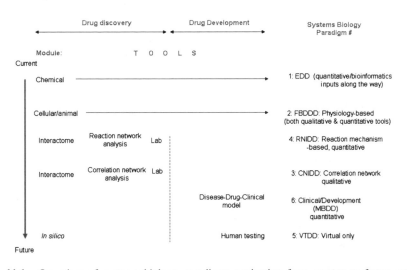

Fig. 11.1 Overview of systems biology paradigms projection from current to future status. Paradigms 1 and 2 presently in use. Note the *pink arrow* timeline indicates introduction of SB paradigms

hybrid, methodology to systematically study complex biological/biomedical experimental data. These data can span the spectrum of experimental experience from in vitro molecular biology to whole body levels. The characterization of these data includes both qualitative and (preferably) quantitative tools. The impact of this characterization helps us evaluate a functional organization that is dynamic, defining within its scope relationships and interactions among various system components and between living systems and their interaction with the environment. In a clear departure from the mechanism-based reductionist approach, SB will embrace both arms of scientific dogma, reductionist and holistic, by employing a well-defined middle ground. This scope and structure of this middle ground may shift in time as more SB global tools are developed and better extraction of emergent properties is instituted. This effort will allow us to obtain a deeper understanding of disease mechanisms (including the upper level interactions), and derive benefits for expedited drug discovery and improved drug safety and efficacy.

Ozbabacan et al. [7] provided an important guidance for the selection and characterization of potential therapeutic targets identified by systems biology: "Given the complexity of human disease and the importance of selecting the right target to avoid costly late-stage drug development failures, advances in network biology may prove to be integral to the long-term success of the pharmaceutical industry."

Most Pharma companies have been less impressed with the past benefits of investing in SB, particularly in the genomic area and OMICs technologies, and are proceeding with caution. We suggest that one way going forward is they change the present organizational structure, which is based on compartmentalized departments in line with reductionism, and apply a more holistic, collaborative, and multidisciplinary work model to SB. In this new paradigm, Pharma's internal culture will need to include a more strategic global way of thinking through out the organization, including throughout upper and middle management. It is also important to note, however, that these types of changes will likely only come gradually, and what pharma must do so must academia.

11.2 Key Technologies and Tools Needed for Development of Systems Biology/CSB

- Strong experimental base: OMICs
- Data storage, analysis and mining via BI tools
- Network Biology:
  1. Pathway analysis and
  2. Modeling, simulation and interrogation tools.

11.3 Steps in Systems Biology/CSB

- Define a preliminary, data-driven (prior knowledge and data mining), hypothesis of a complex process (note that subsequent iterations will define this step further).
- Collect global, dynamic OMICs data with a multitude of different technologies for healthy and disease phenotypes over a range of environmental and genetic changes (from perturbed experiments)
- Analyze them and confront with available information (mining)
- Identify a network model (by exploring the parameter space) that quantitatively recapitulates prior observations and predicts behavior in new environments, and validate it with data collected: reduce model into a correlation or reaction network
- Simulate and interrogate data *in silico* and discover new targets
- Repeat the cycle (return to data collection in order to improve experimental conditions, etc.) to distinguish between competing model hypotheses; repeat the above steps to refine the model over successive iterations and resolve the model inconsistencies and revisions in the topology and in regulatory circuits

The above is strictly an academic flowchart. For a real-life scenario, see below Flowchart (Cookbook); called Systems Biology paradigm by Michelson (please note this diagram represents a culmination of this review, in terms of practical application of systems approach and SB). A description provided by Schadt et al. [8] is structurally similar to what Fig. 11.2 provides.

Chapter(s)

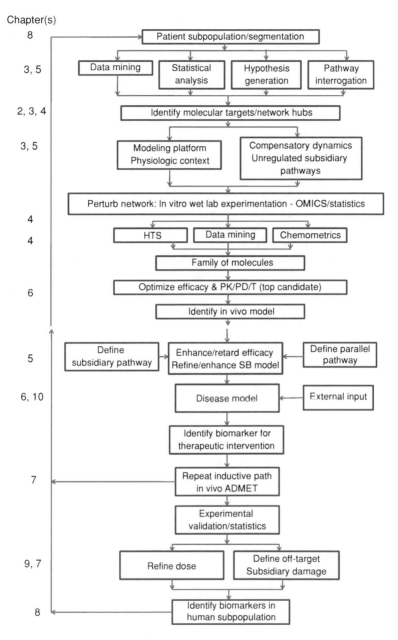

Fig. 11.2 Systems biology paradigm (Flow Chart) by Michelson. A practical guide (flowchart) for Systems Biology discovery (step-by-step)

The following are detailed comments on Fig. 11.2. Remember, that Pharma needs to not only find a molecule that works (efficacy) but that gets to the right place in the right concentration for a long enough period (PK/PD) without causing subsidiary damage (Tox). Then they must determine in which subpopulation of humans those criteria are best met (biomarkers). This practical workflow (as captured in the flowchart in the Fig. 11.2) can be repeated as a learning feedback loop returning to the beginning as necessary.

Identifying and characterizing the patient subpopulation identification and its segmentation is a first step. The key is to use either deductive logic (from data mining and/or statistical analysis) or inductive logic (hypothesizing as to the extent and impact of a pathway), identify a candidate molecular target for a particular kind of intervention. Then assure yourself of its potential viability both with a modeling platform of the physiological context (as complete as possible) that will identify possible (hypothetical) compensatory dynamics (e.g., unregulation of subsidiary pathways, etc.). Then, using a highly focused wet lab experimentation strategy, explore the hypothetical landscape (at any level that is feasible, but typically in vitro). Then, when you think you understand your target "well enough", see if there is any molecule that will manipulate it to therapeutic advantage. This can be done using on HTS machinery and robots, in silico data mining and chemometrics, or both simultaneously. Once one has a family of molecules (post screening—whether *in silico* or HTS), the leads are typically "optimized" by medicinal chemistry to insure "optimal" efficacy (i.e., are we better—more potent, less toxic—than our competitor?) and that exhibit "optimal" PK/PD/Tox profiles to "insure" one is minimizing the chance of failure in Clinical Development. This is typically an iterative process with close interactions between the in vivo biologists and the medicinal chemists, but the output of this "conversation" is typically the top candidate compound and a few back-ups (just in case). Once one has good reason to believe that there are a family of molecules (it is NEVER just one) that might "work" in a more completely physiological environment (i.e. in vivo), find an appropriate in vivo model (typically a lower species like rodent) that you can make "sick" and then make better. Then, to close the loop to SB even tighter, refine/enhance your model to uncover subsidiary dynamics that might indicate subsidiary or parallel pathways that could either enhance or, more likely, retard the efficacy of your candidate molecule. Then, use the model to develop, refine, and posit more complete testable hypotheses (feedback back to induction again), and design the right experiment in the right way to yield the most informative insight as to how the biology and chemistry are interacting. The work product of this effort should yield some level of insight into the dynamics underlying the disease process and its response to external manipulation. That knowledge should, in turn, yield insights into the existence and accessibility of any informative biomarkers one might use to design/optimize a clinical trial and to eventually identify a subpopulation that will likely form the best subject cohort for therapeutic intervention. Now, do it again for ADMET. This loop of scientific induction/deduction and focused hypothesis generation and experimental validation forms the Holy Grail of DDD.

11.4 Benefits of Systems Biology and CSB

- Cut R&D time and costs
- Improve drug target identification
- Improve drug target validation
- Improve the quality of lead compounds
- Optimize lead prioritization
- Develop targets for combination drug therapies
- Identify biomarkets for:

  1. DD and validation
  2. Toxicity
  3. Clinical trials

- Improve ADMET prediction
- Minimize failures due to toxicity
- Optimize strategy and efficiency of clinical trials
- Broaden drug indications in post-market development
- Improve diagnostics for complex diseases
- Target drugs to market segmentation (personalized medicine)
- Guide bioprocess engineering

Finally, we present Table 11.1 (modified from [9]) which summarizes different mathematical tools of BI and SB as relevant to biomedical R&D. Note that not all tools are mentioned in this monograph.

11.5 Summary

- The Achilles heel of the present approach is a strong emphasis on high-affinity ligands.
- The full exploration of biochemical networks will dramatically change the drug discovery process.
- A key systems biology tool, reconstruction of biological networks, represents an emerging field, undergoing explosive expansion; it will enable efficient mapping of genes onto function.
- Qualitative reconstruction of pathways may generate enough information for lead discovery, rather than the current approach of attempting to build fully detailed kinetic models.
- Computational Systems Biology (CSB) is rapidly evolving and will generate very important therapeutically-significant targets soon.
- *In silico* pharmacology is only in a rudimentary state, but will be important for clinical MBDD development.

Table 11.1 Computational tools of SB and BI (adapted from Materi et al. [9])

Method	Description	Advantages	Disadvantages	References
Deterministic/dynamic				
Difference equations	Discrete time and continuous space	Approximation of ODE	Less common	
Ordinary differential equations (ODE)	Numerical solution of reaction-rate equations coupled to mass balance	The most common formalism Deterministic; robust solution	Temporal modeling only Uniformity in space and cocentration	
Partial differential equations (PDE)	Spatial and temporal dependence through partial derivatives; numerical solution	Well-understood formalism Time and space coordinates	Cannot model discontinuous transitions	
Maps				
Discrete time	Parameterized by a discrete-time	Suited for dynamic systems	Approximated by discrete quantities	
Continuous time	Parameterized by a continuous-time		Approximated by discrete event system	
Stochastic processes (random dynamical systems)				
Random mapping (probablity distribution) between an initial state and a final state				
Non-Markovian	Generalized master equation featuring continuous time with memory of past events	Employment of variety of time events	Transitions between states are discrete	
Markovian			Non deterministic	
Series of reaction rate equations solved by master reaction and random number generator			Spatial and temporal modeling tool	
Jump Markovian	Monte Carlo type	Continuous time with no memory of past events		
Continuous Markovian	Stochastic differential equations or a Fokker-Planck equation	Continuous time, continuous state space, events occur continuously		
Gillespie algorithm	Dynamic Monte Carlo method or similar to the kinetic Monte Carlo	The most frequent algorithm		
Petri nets	Place/transition net Description of discrete distributed systems	Robust, suited for complex systems Non mathematical Both quantitative and qualitative	Limited to tempotal processes Often a linear approximation implemented	

(continued)

Table 11.1 (continued)

Method	Description	Advantages	Disadvantages	References
Agent-based methods (ABM)	Objects treated as intelligent agents	Non mathematical	Computationally costly for large number of objects; less developed for ODEs and PDEs	
	Agents follow predetermined rules to cross multiple spatiotemporal scales	Both qualitative and quantitative		
Bayesian networks	Probabilitic graphical model represented by a set of random variables and their conditional interdependencies	Modeling of time and spatial processes Efficient algorithm	No simple conversion from rate constants Requires to speicify a prior distribution for all unknown parameters	
Phylogenetics tools: Trees	A graph theory with different graph concepts	Suitable for sequence variables (protein seuences)	It typically involves high-dimensional integrals Only some tree concepts are relevant to biology	
Cellular automata (CA)	A discrete model in computational theory	Non mathematical		
Boolean networks	A set of Boolean variables whose state is determined by other variables	Non mathematical Provides rules for input/output functions		
Lattice models	A branch of Boolean logic	Suited for system topology and biological process dynamics		
Coarse-grained (CG) methods	System is broken down into small parts, either the system itself or its description CG description considers large subcomponents	For investigating the longer time- and length-scale dynamics	Atomistic description is neglected	
Cellular automaton	Regular grid of cells with finite number of states	Genetic algorithm is a special tupe of CA		

(continued)

Table 11.1 (continued)

Method	Description	Advantages	Disadvantages	References
DCA algorithm	Algorithm design paradigm based on multi-branched recursion	Provide both quantitative and qualitative outputs		
Tesseletion	Nearest neighbor and Voronoi domain analyses	Allows for spatial 3-D morphology and differentiation		
Newer computiong tools				
Stochastic Pi calculus	Language for concurrent processes	Simple description of concurrent processes	Less developed	
	Synchronized input and output	Structured as grapgs		
Graph-theoretic methods (hypergraphs)	A graph partitioning method	Suitable for parallel computing software	Less developed	
Fuzzy methods	A cellular neural network governed by fuzzy local rules	Suitable for control logic and artificial intelligence	Less developed	
Mean-field theory	Model that neglects correlations between interacting entities	Frequently applied in physics	Less developed	
		For modeling of spatial scales in biology		

- *In silico* PKPD/ADMET and biochemical-mechanistic methods will become the standard approach in the coming few years via the employment of BI and SB tools at the multiscale, whole-body level.
- Identifying and targeting (therapeutically) systems emergent properties is the major goal for coming years. This will cause a paradigm shift in R&D activity in pharma and help the establishment of individualized medicine.
- Drug formulation and delivery is very mature area of the R&D process, while targeting is a rapidly developing tool.
- Process design, optimization and scale up (not covered in this review) is an art rather than a rigorous engineering discipline, although some quantitative methods are available, e.g. in the bioreactor design area and scale-up. Process engineering aspects are important part of the global systems approach.
- *In silico* clinical development will only become a reality after some effort is exerted; some partial solutions (in SB and BI) and tools already exist and the first attempts have been made.

References

1. Kola I (2008) The state of innovation in drug development. Clin Pharmacol Ther 83(2):227–230
2. Nicholson JK (2006) Global systems biology, personalized medicine and molecular epidemiology. Mol Syst Biol 2:52
3. Grimaldi D, Claessens YE, Mira JP, Chiche JD (2009) Beyond clinical phenotype: the biologic integratome. Crit Care Med 37(1 Suppl):S38–S49
4. Hood L, Heath JR, Phelps ME, Lin B (2004) Systems biology and new technologies enable predictive and preventative medicine. Science 306(5696):640–643
5. Van der Greef J, McBurney RN (2005) Innovation: Rescuing drug discovery: in vivo systems pathology and systems pharmacology. Nat Rev Drug Discov 4(12):961–967
6. Ridgway D, Broderick G, Ellison MJ (2006) Accommodating space, time and randomness in network simulation. Curr Opin Biotechnol 17(5):493–498
7. Ozbabacan SEA, Gursoy A, Keskin O, Nussinov R (2010) Conformational ensembles, signal transduction and residue hot spots: Application to drug discovery. Curr Opin Drug Discov Develop 13(5):527–537
8. Schadt EE, Friend SH, Shaywitz DA (2009) A network view of disease and compound screening. Nat Rev Drug Discov 8(4):286–295
9. Materi W, Wishart DS (2007) Computational systems biology in drug discovery and development: methods and applications. Drug Discovery Today 12(7-8)295–303